A New Approach to Science: An exploration that fuses science and religion to perceive, interpret and transform reality

Contents

BIOGRAPHIC NOTE ABOUT THE AUTHOR 9

PREFACE 11

FOREWORD 13

I   INTRODUCTION 14

       Classes and Categories are the Foundation of Philosophy 20

       Methodology Followed in the Construction of the "New Approach to Science" 24

       Powers: thought, reason and expression 27

II   SCIENCE IS LOVE 33

III   THE EXPLORATORY METHOD OF SCIENCE 45

       PHILOSOPHICAL ARGUMENT 47

          Mercy 47

          Wisdom 50

          Power 50

          Situations with Challenges, Problems, Tests and Difficulties 53

          Sources of Wisdom 54

          Intelligence 55

- Human Kingdom and other Kingdoms in Nature 56
- Stages and Stations 57
- Relationships 59
- Laws 60
- Parameters and Indicators of Progress 60
- Applying the Single Science to the Cause of Maturity 62

OTHER HUMAN FACULTIES LINKED TO THE MATURITY CAUSE 65

- Senses 65
- Discernment and Self-control 65
- Tenderness, Compassion and Empathy 67
- Two Natures 70
- The Heart's Role in the Decision Making Process Must also be Considered 71
- Free Will 72
- Sense of our Own Powerlessness when we Realize the Unfathomable Depths of His Wisdom in the Books of Revelation and Creation 73

SUMMARY OF THE FOUR CAUSES TAUGHT BY THE GREEK PHILOSOPHERS 75

IV   THE FORMATIVE CAUSE AND THE FORMATIVE METHOD OF SCIENCE 78

PHILOSOPHICAL ARGUMENT 79

General Properties of Matter 83

Religious and Scientific Principles 84

Love 86

The Soul 88

Relationships 90

Parameters and Indicators 90

Applying the Single Science to the Formative Cause 90

OTHER HUMAN FACULTIES LINKED TO THE FORMATIVE CAUSE . 95

Hormones and Emotions 95

Power of Imagination 95

Sense of Religion 96

Sense of Justice 96

V   THE FINAL CAUSE AND THE EXPLICATIVE METHOD OF SCIENCE 101

PHILOSOPHICAL ARGUMENT 102

Relationships 114

Parameters and Indicators of Protection, Equality, Prevention and Security 114

    Applying the Single Science to the Final Cause 114

  OTHER HUMAN FACULTIES LINKED TO THE FINAL CAUSE 121

    Sense of Obligation also Called Sense of Responsibility 122

    The Senses of Fear and Shame 123

    Memory 126

VI THE MATERIAL CAUSE AND THE QUALITATIVE METHOD OF SCIENCE 131

  PHILOSOPHICAL ARGUMENT: 132

    Are Spiritual Qualities Appreciated as Part of what a Chair is Made of? 135

    Are Convictions and Cultural Values Also Appreciated as Part of What a Chair is Made of? 137

    Supplies 138

    The Specific Properties of Matter 138

    Relationships 139

    Parameters and Indicators 139

    Applying the Single Science to the Material Cause 139

  OTHER HUMAN FACULTIES LINKED TO THE MATERIAL CAUSE .. 144

    Senses 144

Common Faculty 144

Conscience 146

Sense of Spirituality 147

Power of Knowledge... 148

Sense of Appreciation 149

VII THE EFFICIENT CAUSE AND THE EXPERIMENTAL METHOD OF SCIENCE 155

PHILOSOPHICAL ARGUMENT.. 156

The Mind... 157

"To be" and "To do" 158

The Word of God 159

The Arts 161

Relationships 163

Parameters of Efficiency... 164

Applying the Single Science to the Efficient Cause............................. 164

Powers: thought, reason and expression 165

OTHER HUMAN FACULTIES LINKED TO THE EFFICIENT CAUSE.. 168

Power of Thought 168

    Power of Reason 170

    Power of Expression.. 171

VIII   THE MATURITY CAUSE AND THE PROPOSITIONAL METHOD OF SCIENCE 176

    PHILOSOPHICAL ARGUMENT 177

        Intelligence 177

        Discernment, Self-control and Volition................................ 177

        Patience.. 178

        Human Kingdom 178

        Kingdoms in Nature 178

        Stages and Stations... 179

        Wisdom... 179

        Mechanism of Control... 179

        Parameters and Indicators of Progress 179

        Evolution 180

    The Quintessence of Knowledge 183

IX   THE FARMERS' SITUATION 186

    The farmers' situation in the context of the exploratory phase of the cycle of scientific research.. 186

The farmers' situation in the context of the formative phase of the cycle of scientific research 195

The farmers' situation in the context of the explicative phase of the cycle of scientific research 206

The farmers' situation in the context of qualitative phase of the cycle of scientific research 216

The farmers' situation in the context of the descriptive and experimental phase of the cycle of scientific research 221

The farmers' situation in the context of the propositional phase of the cycle of scientific research 227

X   WORKS CITED 235

BIOGRAPHIC NOTE ABOUT THE AUTHOR

The author worked as a professor in Colombia for FUNDAEC's University for Rural Wellbeing (Non-Profit Foundation for the Application and Teaching of Science) and the Javeriana University.

His prior publications include: *En Síntesis la Ciencia es Amor* (1992) and *En la Encrucijada: Una Nueva Perspectiva. En Honor a las Mujeres del Mundo* (1999), which Dr. Jairo Roldan, PhD in Philosophy of Science from the Sorbonne University and one of Colombia's foremost physicists praised as expounding a novel approach to science and governance. Dr. Roldan said:

> The third chapter ... constitutes the core of the conceptual framework of the book, framework is based on Aristotle's four causes. The author conducts a personal and original interpretation of these difficult topics in order to find the way towards a synthesis of knowledge and apply it to the work of grassroots organizations. The most original approach is an attempt to link the four causes with different powers, laws, properties, principles and values, referring to God, nature and the human being, and the effort to transform a highly abstract topic into a tool useful for practical problems.
>
> It is in this chapter also where the author presents another of his main theses according to which science is love. Inspired by Plato's dialogue 'The banquet', where the great philosopher speaks of 'the science of what is beautiful', and the writings of Abdu'l-Bahá, where love is identified as the great power that keeps together the universe, professor Duque identifies science with love. The question would love to what? It is not difficult to imagine that a response would be: love of beauty. However, it is possible to try an identification of beauty and

truth, in a mystical sense, identify the real science with a search of the attributes of the Creator, among them the Beauty. Being the Creator a Unique Reality, love for His beauty, for His reality, it would be one thing. To have Religion as purpose the pursuit of spiritual truths, and to become reality one, both Science and Religion would fundamentally seek the same. From there the essential harmony between them, which is one of the fundamental principles which professor Duque accepts. His thesis, as can be seen, is inspirational for innovative research about the nature of science.

The fifth chapter deals with the second great topic of the book: the art of governing. The author submits that to rule it is necessary to first discern between all the possible scenarios the more desirable, in other words that should be an option. Beginning with an exercise whereby some parameters are investigated to measure development, Leonardo Duque presents his reflections, based on his extensive experience as a member of the administrative order of the Bahá'i Faith, about democracy, the participation, the legitimacy of a Government, governance and power.[1]

He was a member of the National Spiritual Assemblies of the Bahá'ís of Costa Rica (1979-1981) and Colombia (1983-2006).

---

[1] Trans. by the author

PREFACE

This book enriches the four causes of Aristotelian thought—i.e., the material, formative, efficient and final causes (Duque, 1.992) – and a previously suggested fifth cause (Duque, 1.999), the Maturity Cause, using the same methodology discovered while working with the four causes.

Each of the five causes is associated with one or two methods of science and to a set of human faculties. The four causes taught by the Greeks are associated with the qualitative, the formative method, the experimental, the explicative method, and the exploratory and propositional methods that emerge from the Maturity Cause. The full conceptual framework based on the five causes is a useful tool for solving every day common issues as well as for resolving large individual and collective challenges.

The outcome of a New Approach to Science, which fuses notions from the Book of Creation and the Book of Revelation in a conceptual framework to perceive, interpret and transform reality, profoundly questions how science is conceived today.

What would be the synergic effect when "science and religion, the two most potent forces in human life"[2] become reconciled?

The author's other books have been dedicated to the relentlessly persecuted Bahá'ís of Iran. This book is also dedicated to them and to the farmers of the world. When a New Approach to Science is used to

---

[2] Shoghi Effendi in the Introduction of a book by Bahá'u'lláh, The Proclamation of Bahá'u'lláh, p. xi.

research the harsh farmers' situation, we find that the human race has hope.

I specially want to thank Marsha Huitt for her highly professional editing recommendations. I also want to thank Professor William Huitt for his scholar comments and those very accurate philosophical questions from Ian Kluge, both of which, for the first time, led me to uncover the methodology that I have been using; which was progressively ingrained deep in my mind. There was never an intentional attempt to keep it in hiding.

My eternal gratitude to my wife and our dear children for their support, patience and love.

FOREWORD

A hallmark of the age in which we live is the freedom to investigate reality and truth. Indeed, beyond the common concept of "freedom", which implies the possibility of a "take-it-or-leave-it" attitude toward a search for truth, it is indeed the *responsibility* of each individual to investigate reality. Search is part of our essential being; it is not an option. The current work explores the paths to truth and the ways that knowledge of causes was structured among the ancient Greeks, applying this to our current age of investigation. Still, the past will necessarily be inadequate to describe the present, much less to visualize a path into the future. An additional ingredient is required and is defined by the current age itself. In the age of the maturity of mankind, it is maturity that complements and brings to fruition the wisdom of past ages. If science is the study of cause and effect in the material world, then the causes of phenomena occur at multiple levels: as the raw substance of material reality, as the form that material reality takes, as the action or influence that shapes material reality, and as the purpose or intention behind a given expression of material reality. When put in the context of human decisions, man's spiritual reality and social context, the maturity engendered in light of the Revelation of Bahá'u'lláh guides this process of knowledge toward a more just and peaceful society. The current study investigates the implications of bringing such maturity to bear on the current state of agriculture and smallholder farmers that are prominent in the so-called developing world. It is another contribution to the search after knowledge that the Bahá'í Revelation has provoked in the collective life of humanity.

Dr. Stephen Beebe - a senior bean researcher at the International Center for Tropical Agriculture (CIAT)

I INTRODUCTION

Harmony between science and religion is one of the principles of the Bahá'í Faith. For a long time they have been considered incompatible. Being aware that prejudices and ideologies are the cause of bias that distort the truth, and believing that "Truthfulness is the foundation of all human virtues. Without truthfulness, progress and success, in all the worlds of God, are impossible for any soul. When this holy attribute is established in man, all the divine qualities will also be acquired"[3], I sought to explore the harmony of science and religion by scrutinizing the four causes taught by the ancient Greek philosophers, which are explained more fully below. I did this for the three reasons that are set forth below:

The first reason is set forth in the following series of quotes by Roger Walsh, from his "*Staying Alive: the psychology of human survival*":

> Our current crises are seen as expressions of the mistaken desires, fears and perceptions that arise from our mistaken identity. Since all the major threats to human survival and wellbeing are human caused, they are of course, deeply, though not exclusively, psychological in origin. The state of the world, then, is a reflection of the state of our individual and collective minds. He points out psychological, educational and sociological strategies that may mitigate our situation.
>
> World issues proceed, as a last resort, from human minds and from the actions that these minds unleash; because of this, the efforts to subdue our fears and egoisms; to overcome people's mistaken

---

[3] 'Abdu'l-Bahá qtd. by Shoghi Effendi, The Advent of Divine Justice, p. 26.

beliefs or transcend non-solidary attitudes are keys in the planetary evolution.

Inasmuch as the state of the world reflects the state of our minds, then what we have called our global "problems" are actually global "symptoms": symptoms of our individual and collective pathologies and immaturities. Therefore, truly curative responses will need to address these internal sources. We will need multifaceted interventions in which we not only try, for example, to feed the hungry and work for peace, but also attempt to address the psychological roots of these problems: the greed, hatred and delusion, and lack of love, compassion, and wisdom, which created them in the first place. We are in a race between consciousness and catastrophe.[4]

The perennial philosophy, which lies at the heart of the great religions and is increasingly said to represent their deepest thinking, suggests a very different view. It views consciousness as central and its development as the primary goal of existence. This development is said to culminate in the condition variously known in different traditions as enlightenment, liberation, salvation ... .

The descriptions of this condition show remarkable similarities across cultures and centuries. Its essence is said to be the recognition that the distortions of our usual state of mind are such that we have been suffering from a case of mistaken identity.[5]

---

[4] Walsh, Staying Alive.

[5] Walsh, El Compromiso con el Planeta.

The second reason is that the Greek philosophers have been, throughout history, universally recognized for their great wisdom, and their books have been amply disseminated.

Third, and most importantly, is that Bahá'u'lláh, when referring to the Greek philosophers, said:

> Although it is recognized that the contemporary men of learning are highly qualified in philosophy, arts and crafts, yet were anyone to observe with a discriminating eye he would readily comprehend that most of this knowledge hath been acquired from the sages of the past, for it is they who have laid the foundation of philosophy, reared its structure and reinforced its pillars. Thus doth thy Lord, the Ancient of Days, inform thee. The sages aforetime acquired their knowledge from the Prophets, inasmuch as the latter were the Exponents of divine philosophy and the Revealers of heavenly mysteries.
>
> Consider Hippocrates, the physician. He was one of the eminent philosophers who believed in God and acknowledged His sovereignty. After him came Socrates who was indeed wise, accomplished and righteous. … He is the most distinguished of all philosophers and was highly versed in wisdom.
>
> After Socrates came the divine Plato who was a pupil of the former and occupied the chair of philosophy as his successor. He acknowledged his belief in God and in His signs which pervade all that hath been and shall be. Then came Aristotle, the well-known man of knowledge. He it is who discovered the power of gaseous matter. These men who stand out as leaders of the people and are pre-

eminent among them, one and all acknowledged their belief in the immortal Being Who holdeth in His grasp the reins of all sciences.[6]

The great Greek philosophers based their explanation of all phenomena on four fundamental "causes", or explanations for phenomena: the Material Cause, the Formative Cause, the Efficient Cause and the final cause. Today's academic world seems uninterested in the significance of the four causes taught by the Greek philosophers. Academics seem apprehensive and uncertain regarding the possibility of unifying knowledge.

Consider, for example, Mikael Stenmark's conception of science:

> The idea seems to be that science should be free from not merely ideological or religious values but also ideological or religious beliefs. I shall therefore often talk not merely about values but also about ideologies and religions. With this clarification in mind, I propose that we define the value-free view of science more precisely in this way: The value-free view of science is the standpoint that science should be autonomous, neutral, impartial, non-responsible, and non-normative.[7]

In supporting his perspective about an impartial science, Stenmark says: "Science should be impartial in the sense that it should not presuppose the truth of any particular political vision, religion or ideology in the validation of scientific theories."[8]

---

[6] Tablets of Bahá'u'lláh Revealed After the Kitáb-i-Aqdas, Tablet of Wisdom, pp. 146-47.

[7] Mikael Stenmark, qtd. in LeRon Shults (ed.), The Evolution of Rationality, p. 51.

[8] ibid. p. 51.

In reference to Stenmark's statement, I would not waste my time trying to demonstrate that God's teachings are false; I would dedicate my ephemeral life to understanding their validity.

It has been said that "our current crises are seen as expressions of the mistaken desires, fears and perceptions that arise from our mistaken identity," and that "the current threats to human survival and well-being *are actually symptoms*, symptoms of our individual and collective mind set. The state of the world is therefore a creation and expression of our own minds, and it is to our own minds that we must look for solutions."[9]

The following questions, among many others, arise from these reflections:

What am I?

Why do I exist? For what do I exist?

How do I envision myself?

Who am I?

Which options do I have once I reach the age of maturity?

To examine the approach to answering these questions, I will start the philosophical discussion of the causes with two quotes by 'Abdu'l-Bahá that serve as the foundation for this search:

> Essential pre-existence is an existence which is not preceded by a cause; essential origination is preceded by a cause. Temporal pre-existence has no beginning; temporal origination has both a beginning and an end. For the existence of each and every thing depends upon four causes: the Efficient Cause, the Material Cause, the formal cause, and the final cause. So this chair has a creator who is a carpenter, a matter which is wood, a form which is that of a chair,

---

[9] Roger Walsh, Staying Alive: the psychology of human survival.

and a purpose which is to serve as a seat. Therefore, this chair is essentially originated, for it is preceded by, and its existence is conditioned upon, a cause. This is called essential or intrinsic origination.[10]

'Abdu'l-Bahá is reported to have said:

Someone desires an explanation of the terms soul, mind and spirit. The terminology of ancient and modern philosophers differs. According to the great ancient philosophers the words soul, mind and spirit implied the underlying principles of life; the essence was expressed under different names and these three terms designated the various functions of the absolute reality, or the operations of the one single essence; for instance, when they dealt with the *sensations of emotion* they called it the *soul*; when they desired to express that *power which discovers the reality of phenomena* they gave it the appellation of *mind* and when they discussed the *consciousness* which pervades the world of creation they gave it the title of *spirit*.[11]

I was not aware of the epistemological discussions about *understanding* that would allow one to say that one has *knowledge* of something, nor had I read about the metaphysical search for the meaning of *truth* when I started my research, and I also did not receive any guidance in how to comprehend the philosophical questions about the meanings of *causes* and *principles*. I believe that, had I had these things, I likely would have lost the path!

---

[10] 'Abdu'l-Bahá, Some Answered Questions, pp.155 -56.

[11] 'Abdu'l-Bahá, Divine Philosophy, p. 119. Emphasis added.

The assumption upon which I based my investigation started believing deeply in my heart that it was possible to bring about harmony between science and religion. The problem today is the fragmentation of knowledge into scientific disciplines, and separation of science and religion in two separate domains. In the "New approach to Science" religion and all disciplines of science are fused into a common conceptual framework: a single science.

I decided to search for a better way to understand reality, believing deeply in my heart that it was possible to bring about harmony between science and religion. I started the puzzle putting a few notions next to each of the four causes that the Greek philosophers taught, and then searching their writings for a meaningful connection. It was frustrating, but ultimately rewarding with patience and perseverance.

Each of the five causes, the four original causes taught by the Greek philosophers and the fifth cause, the Maturity Cause, that I propose, are associated with one or two methods of science and to a set of human faculties. The Formative Cause to the formative method; the Material Cause to the qualitative method; the Efficient Cause to the descriptive and experimental methods, and the final cause, taught by the Greeks, to the explicative method. The exploratory and propositional methods emerge from the Maturity Cause, which also includes the concept of cycle and its phases and the importance of taking into account all the kingdoms of creation while having to choose among different options.

CLASSES AND CATEGORIES ARE THE FOUNDATION OF PHILOSOPHY

Classes: clusters, groups, categories, sets, conglomerates and the answer to: "What type of object?" "What species of plants, or insects?"

Socrates explains why categories should be included in the Efficient Cause, which we will study later on in this document:

Stranger- Then, not to exclude anyone who has ever speculated at all upon the nature of being, let us put our questions to them as well as to our former friends.

Theaetetus- What questions?

Stranger- Shall we refuse to attribute being to motion and rest, or anything to anything, and assume that they do not mingle, and are incapable of participating in one another? Or shall we gather all into one class of things communicable with one another? Or are some things communicable and others not? Which of these alternatives, Theaetetus, will they prefer?

Theaetetus- I have nothing to answer on their behalf. Suppose that you take all these hypotheses in turn, and see what are the consequences which follow from each of them. ...

Stranger- And now, if we suppose that all things have the power of communion with one another -what will follow?

Theaetetus- Even I can solve that riddle.

Stranger- How?

Theaetetus- Why, because motion itself would be at rest, and rest again in motion, if they could be attributed to one another.

Stranger- But this is utterly impossible.

Theaetetus- Of course.

Stranger- Then only the third hypothesis remains.

Theaetetus- True.

Stranger- For, surely, either all things have communion with all; or nothing with any other thing; or some things communicate with some things and others not.

Theaetetus- Certainly.

Stranger- And two out of these three suppositions have been found to be impossible.

Theaetetus- Yes.

Stranger-Every one then, who desires to answer truly, will adopt the third and remaining hypothesis of the communion of some with some.

Theaetetus- Quite true.

Stranger- This communion of some with some may be illustrated by the case of letters; for some letters do not fit each other, while others do.

Theaetetus- Of course.

Stranger- And the vowels, especially, are a sort of bond which pervades all the other letters, so that without a vowel one consonant cannot be joined to another.

Theaetetus- True.

Stranger- But does every one know what letters will unite with what? Or is art required in order to do so?

Theaetetus- What is required.

Stranger- What art?

Theaetetus- The art of grammar.

Stranger- And is not this also true of sounds high and low?-Is not he who has the art to know what sounds mingle, a musician, and he who is ignorant, not a musician?

Theaetetus- Yes.

Stranger- And we shall find this to be generally true of art or the absence of art.

Theaetetus- Of course.

Stranger- And as classes are admitted by us in like manner to be some of them capable and others incapable of intermixture, must not he who would rightly show what kinds will unite and what will not, proceed by the help of science in the path of argument? And will he not ask if the connecting links are universal, and so capable of intermixture with all things; and again, in divisions, whether there are not other universal classes, which make them possible?

Theaetetus- To be sure he will require science, and, if I am not mistaken, the very greatest of all sciences.

Stranger- How are we to call it? By Zeus, have we not lighted unwittingly upon our free and noble science, and in looking for the Sophist have we not entertained the philosopher unawares?

Theaetetus- What do you mean?

Stranger- Should we not say that the division according to classes, which neither makes the same other, nor makes other the same, is the business of the dialectical science?

Theaetetus- That is what we should say.

Stranger- Then, surely, he who can divide rightly is able to see clearly one form pervading a scattered multitude, and many different forms contained under one higher form; and again, one form knit together into a single whole and pervading many such wholes, and many forms, existing only in separation and isolation. This is the knowledge of classes which determines where they can have communion with one another and where not.

Theaetetus- Quite true.[12]

Including "categories" as part of this proposal allows it to be perfected by others to higher levels of excellence.

---

[12] Plato, Sophist.

As the reader can perceive, in order to suggest a new concept of science, the author has grouped a carefully chosen set of notions to each cause. Of course that in a search for a harmonious relation between science and religion the task implied to explore for those notions in both realms: the Book of Creation and the Book of Revelation. The novelty of the way of the adopted grouping is that each cause has a conceptual framework that includes a principle that encompasses all creation, concepts closely related to that principle, certain general laws, a structure of relationships, a method, a set of human faculties, some moral values and the associated parameters and indicators.

> Man is said to be the greatest representative of God, and he is the Book of Creation because all the mysteries of beings exist in him. If he comes under the shadow of the True Educator and is rightly trained, he becomes the essence of essences, the light of lights, the spirit of spirits; he becomes the center of the divine appearances, the source of spiritual qualities, the rising-place of heavenly lights, and the receptacle of divine inspirations. If he is deprived of this education, he becomes the manifestation of satanic qualities, the sum of animal vices, and the source of all dark conditions. [13]

METHODOLOGY FOLLOWED IN THE CONSTRUCTION OF THE NEW APPROACH TO SCIENCE

The single science I was looking for is the intellectual capacity to classify any notion relevant to the solution of a problem in one of the five causes (please refer to the table below), in a loving and systematic search for the beauty of truth, going through the phases of the cycle of scientific research.

---

[13] Abdu'l-Baha, Some Answered Questions, p. 236.

Please refer to the table below showing the methodology followed in the construction of the "New Approach to Science" by inter-weaving the causes horizontally and vertically. Rows 7 and 8 suggest possible outcomes when applying the conceptual framework.

This methodology unfolded gradually during the research process.

	FINAL CAUSE:	MATERIAL CAUSE:	FORMATIVE CAUSE:	EFFICIENT CAUSE:	MATURITY CAUSE	
1	**Question:** Why? For what?	**Question:** With what?	**Question:** How is the arrangement?	**Questions:** With what being? With whom? With what type?	**Question:** With which option? when addressing problems, difficulties, issues and needs	**CAUSE OF MATURITY**
2	**Essence:** the power of law itself, i.e., that which is a condition to a certain aspect of life or nature, and explains its purpose, its mission	**Essence:** The Spirit. Elements, substances and raw materials. Specific properties of matter, peculiarities of things and characteristics. Moral values, also called spiritual values. Knowledge, beliefs and social values	**Essence:** Love and soul as its power. **The general properties of matter:** form, size, mass, temperature, movement and position in time and space **Hormones, feelings and emotions**	**Essence:** The Word of God. **To be:** animated and unanimated beings. **Forces:** power of mind, energy, capacities, potential, talents, vocations, arguments, concepts. **Categories:** genres, clusters, groups. Interacting entities within a system or a subsystem.	**Essence:** Mercy. **Sources of Wisdom:** What history or experience, the diverse disciplines of science and the world religions say about the issue to be addressed **Kingdoms, Cycles and phases**	**MATERIAL CAUSE**
3	Social order, laws and norms; Natural laws; Religious laws and ordinances. Pacts, agreements and covenants. Individual and collective commitment to ideals	**Spiritual laws**, the **standards** that regulate quality and **laws of possession** **Goals** and objectives	**Laws** related to unity, harmony, justice and the law of gravity. **Goal:** Aspirations and desires to reach the expected vision or design	**Laws** of thermodynamics, labor **law** and tools **regulations**. Grammar **rules**. The Word of God is **Law** **Fruits of labor:** results, harvest, services, products and leftover materials or efforts	**Laws:** related to the inherent responsibilities of the assumed decision taking into account the effects in the other Kingdoms and in the Human Kingdom. **Goal:** Stewardship in the management of the trust	**FINAL CAUSE**

Goal: the end, the mission, the purpose

Powers: Memory and senses of responsibility, fear and shame **Advice** from those with experience. The **consequences** of obeying and disobeying.	**Powers:** conscience, knowledge, common faculty and senses of spirituality and appreciation **Specific lines of action or policies** to consolidate ethics, knowledge, quality and efficacy	**Powers:** imagination and senses of justice and religion **The practical application of the principles related to the general properties of matter** in relation to the desire organization, arrangement and design	**Powers:** thought, reason and expression **To do:** movements: methods, processes, activities, abilities, skills, arts, technologies, mechanisms.	**Powers:** Intelligence, patience, self-control, discernment, compassion, empathy, volition **Decisions being made at the individual and collective level:** Governmental Institutions, Administration of private enterprises, NGOs, families or individuals. The human decisions in relation to the stages of a plan or phases of a cycle **The spontaneous differential response** to the stimuli received by the different kingdoms **The effects of the decisions made by humans in all kingdom**s **The developmental stage** of the individual, plant or animal **Monitoring and evaluating** the strategy and its stages and the cycle and its phases

4

EFFICIENT CAUSE

27

5	Situations of risk and danger. **Relationships** of gratitude, loyalty, reciprocity, mutuality, of fear and protection; and also guilt, repentance, punishment and reward.	**Relationships** of belonging, possession, distinction and characterization	**Religious and Scientific principles** related to the general properties of matter	**Relations of:** Cause and effect, logic and reason.	Situations in which the human kingdom is being challenged to choose or, in the case of the mineral, plant and animal kingdoms, the spontaneous differential response to the stimuli received. **Relationships:** are those that interconnect the sources of wisdom	**FORMATIVE CAUSE**
6	**Parameters and indicators** of protection, equality, prevention and security in the fulfillment of the mission	**Parameters and indicators** of efficacy and quality in attaining the goals and objectives	**Parameters and indicators** of unity, beauty, symmetry, harmony, reciprocity and justice in reaching the desire vision or design	**Parameters and indicators** of efficiency and productivity in achieving the results	**Parameters and indicators** of **progress**: Such as level of respect for the individual to exercise his (her) free will, level of commitment, number of pledges, and number of recommendations or referrals	**CAUSE OF MATURITY**

7   If a CEO of a factory delivers a product that is safe for the customers; with very good quality; a comfortable and beautiful design that has been reached being fair to his workers (profit sharing); in an Efficient way and at a reasonable cost; and, that is recyclable after a lasting life, she will get a lot of referrals.

8   If an educational program teaches adolescents, to be individually and collectively responsible, to nurture their spiritual endowment, to moderate their impulses and aim towards praiseworthy aspirations, to develop their powers of expression and the skills and capabilities to become protagonist of constructive social change, and to acquire the ability to make sound choices in their lives, will. become highly recommended

The suggested conceptual framework's desired end result is to deepen the bonds between man's inner nature and his material surroundings. Similarly, the conceptual framework expounds, with clear transparency, the importance of education as the foundation for the development of the peoples of the world, and of agriculture as the basis of order and system in society.

I was confronted with the difficult reality of the rural areas while working at FUNDAEC's University for Rural Well-being (Non-Profit Foundation for the Application and Teaching of Science). I fell in love with FUNDAEC's scientific approach to address the needs of the rural areas and I am pleased to dedicate this book to the farmers of the world.

Scientific research can either be a journey with different stations where one can pass through one or some of them, or as a cycle where one goes through all the phases. Both ways are valid. Individuals or groups can begin exploring a problem and after some reflection propose a solution, or follow all the phases shown in the graphic below.

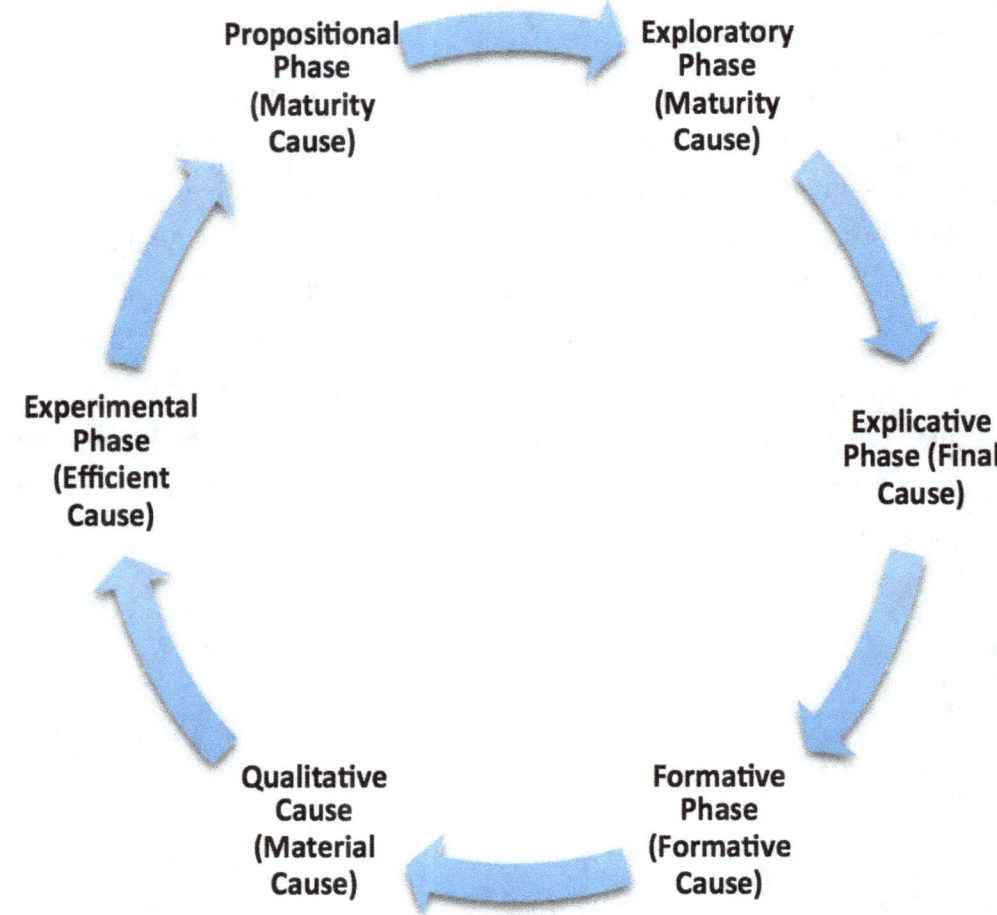

Each phase of the cycle of scientific research is connected to a cause. It will be demonstrated that each cause has its own set of faculties of perception, inherent essence, laws, structure of relationships, method(s), parameters and indicators. The reader can observe in the above table that to each essence, the author has associated a well-articulated set of key notions:

> The essence of Mercy, as the cause of maturity, was the reason that help the author to bring into existence the set of notions of the exploratory and propositional methods of science.

The essence of the Laws of God, as the final cause, put the author in the right track to articulate the set of notions of the explicative method of science.

The essence of the Spirit, as the material cause, was the way to conceive the fitting together of the set of notions of the qualitative method of science.

The essence of Love, as the formative cause, showed the author how to justify the origination of the set of notions of the formative method of science.

The essence of the Word of God, as the efficient cause, was the key used by the author to generate the set of notions of the experimental and descriptive methods of science.

Also, a set of human faculties is proposed for each of the five causes. Because many of the problems that we have are caused by ourselves, it is of extreme importance that we also understand what faculty needs to be further developed. The recognition of such faculties is very important in determining a more efficient way to potentiate them individually or collectively, during the pedagogical process and also in the psychological approach to human behaviors.

In order to reach a holistic perspective, the conceptual framework for each one of the causes is illustrated in a progressive way with two examples: one about a chair and another about the situation of the farmers today.

The full conceptual framework is a useful tool for solving every day common issues as well as resolving complex individual and collective challenges.

Because the exploratory phase is considered the starting point of the cycle and the propositional phase the ending point of the cycle, we will start with an in-depth discussion of the Maturity Cause.

II SCIENCE IS LOVE

> "How can feeble reason encompass the Qur'án,
> Or the spider snare a phoenix in his web?
> Wouldst thou that the mind should not entrap thee?
> Teach it the science of the love of God!"[14]

Consider the concept of science as taught by Socrates's teacher, the wise woman Diotima of Mantineia. The author refers to Diotima's following text, in each one of the causes taught by the Greek Philosophers, to support the conclusion that a single science is love: attraction to the beauty of truth. Socrates, talking to his teacher, says:

> 'What then is Love?' I asked; 'Is he mortal?' 'No.' 'What then?' 'As in the former instance, he is neither mortal nor immortal, but in a mean between the two.' 'What is he, Diotima?' 'He is a great spirit (daimon), and like all spirits he is intermediate between the divine and the mortal.' 'And what,' I said, 'is his power?' 'He interprets,' she replied, 'between gods and men, conveying and taking across to the gods the prayers and sacrifices of men, and to men the commands and replies of the gods; he is the mediator who spans the chasm which divides them, and therefore in him all is bound together, and through him the arts of the prophet and the priest, their sacrifices and mysteries and charms, and all, prophecy and incantation, find their way. For God mingles not with man; but through Love all the intercourse, and converse of God with man, whether awake or asleep, is carried on. The wisdom which understands this is spiritual; all other wisdom, such as that of arts and handicrafts, is mean and vulgar. Now these

---

[14] Bahá'u'lláh, The Four Valleys, p. 52.

spirits or intermediate powers are many and diverse, and one of them is Love. 'And who,' I said, 'was his father, and who his mother?' 'The tale,' she said, 'will take time; nevertheless I will tell you. On the birthday of Aphrodite there was a feast of the gods, at which the god Poros or Plenty, who is the son of Metis or Discretion, was one of the guests. When the feast was over, Penia or Poverty, as the manner is on such occasions, came about the doors to beg. Now Plenty who was the worse for nectar (there was no wine in those days), went into the garden of Zeus and fell into a heavy sleep, and Poverty considering her own straitened circumstances, plotted to have a child by him, and accordingly she lay down at his side and conceived love, who partly because he is naturally a lover of the beautiful, and because Aphrodite is herself beautiful, and also because he was born on her birthday, is her follower and attendant. And as his parentage is, so also are his fortunes. In the first place he is always poor, and anything but tender and fair, as the many imagine him; and he is rough and squalid, and has no shoes, nor a house to dwell in; on the bare earth exposed he lies under the open heaven, in-the streets, or at the doors of houses, taking his rest; and like his mother he is always in distress. Like his father too, whom he also partly resembles, he is always plotting against the fair and good; he is bold, enterprising, strong, a mighty hunter, always weaving some intrigue or other, keen in the pursuit of wisdom, fertile in resources; a philosopher at all times, terrible as an enchanter, sorcerer, sophist. He is by nature neither mortal nor immortal, but alive and flourishing at one moment when he is in plenty, and dead at another moment, and again alive by reason of his father's nature. But that which is always flowing in is always flowing out, and so he is never in want and never in wealth; and, further, he is in a mean between ignorance and knowledge. The truth of the matter is this: No god is a philosopher or seeker after wisdom, for he is wise already; nor does any man who is wise seek after wisdom. Neither do the ignorant seek after Wisdom. For herein is the evil of ignorance, that he who is neither good nor wise is nevertheless satisfied with himself: he has no desire for that of which he feels no want.' 'But-who then, Diotima,' I said, 'are the lovers of wisdom, if they are neither the wise nor the foolish?' 'A child may answer that question,' she replied; 'they are those who are in a mean between the two; Love is one of them. For wisdom is a most beautiful thing, and Love is of the beautiful; and therefore Love is also a philosopher: or lover of wisdom, and being a lover of wisdom is in a mean between the wise and the ignorant. And of this too his birth is the cause; for his father is wealthy and wise, and his mother poor and foolish. Such, my dear Socrates, is the nature of the spirit Love. The error in your conception of him was very natural, and as I imagine from what you say, has arisen out of a confusion of love and the beloved, which made you think that love was all beautiful. For the beloved is the truly beautiful, and delicate, and perfect, and blessed; but the principle of love is of another nature, and is such as I have described.'

I said, 'O thou stranger woman, thou sayest well; but, assuming Love to be such as you say, what is the use of him to men?' 'That, Socrates,' she replied, 'I will attempt to unfold: of his nature and birth I have already spoken; and you acknowledge that love is of the beautiful. But some one will say: Of the beautiful in what, Socrates and Diotima?-or rather let me put the question more dearly, and ask: When a man loves the beautiful, what does he desire?' I answered her 'That the beautiful may be his.' 'Still,' she said, 'the answer suggests a further question: What is given by the possession of beauty?' 'To what you have asked,' I replied, 'I have no answer ready.' 'Then,' she said, 'Let me put the word 'good' in the place of the beautiful, and repeat the question once more: If he who loves good, what is it then that he loves? 'The possession of the good,' I said. 'And what does he gain who possesses the good?' 'Happiness,' I replied; 'there is less difficulty in answering that question.' 'Yes,' she said, 'the happy are made happy by the acquisition of good things. Nor is there any need to ask why a man desires happiness; the answer is already final.' 'You are right.' I said. 'And is this wish and this desire common to all? and do all men always desire their own good, or only some men?-what say you?' 'All men,' I replied; 'the desire is common to all.' 'Why, then,' she rejoined, 'are not all men, Socrates, said to love, but only some them? whereas you say that all men are always loving the same things.' 'I myself wonder,' I said,-why this is.' 'There is nothing to wonder at,' she replied; 'the reason is that one part of love is separated off and receives the name of the whole, but the other parts have other names.' 'Give an illustration,' I said. She answered me as follows: 'There is poetry, which, as you know, is complex; and manifold. All creation or passage of non-being into being is poetry or making, and the processes of all art are creative; and the masters of arts are all poets or makers.' 'Very true.' 'Still,' she said, 'you know that they are not called poets, but have other names; only that portion of the art which is separated off from the rest, and is concerned with music and metre, is termed poetry, and they who possess poetry in this sense of the word are called poets.' 'Very true,' I said. 'And the same holds of love. For you may say generally that all desire of good and happiness is only the great and subtle power of love; but they who are drawn towards him by any other path, whether the path of money-making or gymnastics or philosophy, are not called lovers -the name of the whole is appropriated to those whose affection takes one form only-they alone are said to love, or to be lovers.' 'I dare say,' I replied, 'that you are right.' 'Yes,' she added, 'and you hear people say that lovers are seeking for their other half;

but I say that they are seeking neither for the half of themselves, nor for the whole, unless the half or the whole be also a good. And they will cut off their own hands and feet and cast them away, if they are evil; for they love not what is their own, unless perchance there be some one who calls what belongs to him the good, and what belongs to another the evil. For there is nothing which men love but the good. Is there anything?' 'Certainly, I should say, that there is nothing.' 'Then,' she said, 'the simple truth is, that men love the good.' 'Yes,' I said. 'To which must be added that they love the possession of the good?' 'Yes, that must be added.' 'And not only the possession, but the everlasting possession of the good?' 'That must be added too.' 'Then love,' she said, 'may be described generally as the love of the everlasting possession of the good?' 'That is most true.'

'Then if this be the nature of love, can you tell me further,' she said, 'what is the manner of the pursuit? what are they doing who show all this eagerness and heat which is called love? and what is the object which they have in view? Answer me.' 'Nay, Diotima,' I replied, 'if I had known, I should not have wondered at your wisdom, neither should I have come to learn from you about this very matter.' 'Well,' she said, 'I will teach you:-The object which they have in view is birth in beauty, whether of body or, soul.' 'I do not understand you,' I said; 'the oracle requires an explanation.' 'I will make my meaning dearer,' she replied. 'I mean to say, that all men are bringing to the birth in their bodies and in their souls. There is a certain age at which human nature is desirous of procreation-procreation which must be in beauty and not in deformity; and this procreation is the union of man and woman, and is a divine thing; for conception and generation are an immortal principle in the mortal creature, and in the inharmonious they can never be. But the deformed is always inharmonious with the divine, and the beautiful harmonious. Beauty, then, is the destiny or goddess of parturition who presides at birth, and therefore, when approaching beauty, the conceiving power is propitious, and diffusive, and benign, and begets and bears fruit: at the sight of ugliness she frowns and contracts and has a sense of pain, and turns away, and shrivels up, and not without a pang refrains from conception. And this is the reason why, when the hour of conception arrives, and the teeming nature is full, there is such a flutter and ecstasy about beauty whose approach is the alleviation of the pain of travail. For love, Socrates, is not, as you imagine, the love of the beautiful only.' 'What then?' 'The love of generation and of birth in beauty.' 'Yes,' I said. 'Yes, indeed,' she replied. 'But why of generation?' 'Because to the mortal creature, generation is a sort of eternity and immortality,' she replied; 'and if, as has been already admitted, love is of the everlasting possession of the good, all men will necessarily desire immortality together with good: Wherefore love is of immortality.'

All this she taught me at various times when she spoke of love. And I remember her once saying to me, 'What is the cause, Socrates, of love, and the attendant desire? See you not how all animals, birds, as well as beasts, in their desire of procreation, are in agony when they take the infection of love, which begins with the desire of union; whereto is added the care of offspring, on whose behalf the weakest are ready to battle against the strongest even to the uttermost, and to die for them, and will, let themselves be tormented with hunger or suffer anything in order to maintain their young. Man may be supposed to act thus from reason; but why should animals have these passionate feelings? Can you tell me why?' Again I replied that I did not know. She said to me: 'And do you expect ever to become a master in the art of love, if you do not know this?' 'But I have told you already, Diotima, that my ignorance is the reason why I come to you; for I am conscious that I want a teacher; tell me then the cause of this and of the other mysteries of love.' 'Marvel not,' she said, 'if you believe that love is of the immortal, as we have several times acknowledged; for here again, and on the same principle too, the mortal nature is seeking as far as is possible to be everlasting and immortal: and this is only to be attained by generation, because generation always leaves behind a new existence in the place of the old. Nay even in the life, of the same individual there is succession and not absolute unity: a man is called the same, and yet in the short interval which elapses between youth and age, and in which every animal is said to have life and identity, he is undergoing a perpetual process of loss and reparation-hair, flesh, bones, blood, and the whole body are always changing. Which is true not only of the body, but also of the soul, whose habits, tempers, opinions, desires, pleasures, pains, fears, never remain the same in any one of us, but are always coming and going; and equally true of knowledge, and what is still more surprising to us mortals, not only do the sciences in general spring up and decay, so that in respect of them we are never the same; but each of them individually experiences a like change. For what is implied in the word 'recollection,' but the departure of knowledge, which is ever being forgotten, and is renewed and preserved by recollection, and appears to be the same although in reality new, according to that law of succession by which all mortal things are preserved, not absolutely the same, but by substitution, the old worn-out mortality leaving another new and similar existence behind unlike the divine, which is always the same and not another? And in this way, Socrates, the mortal body, or mortal anything, partakes of immortality; but the immortal in another way. Marvel not then at the love which all men have of their offspring; for that universal love and interest is for the sake of immortality.'

I was astonished at her words, and said: 'Is this really true, O thou wise Diotima?' And she answered with all the authority of an accomplished sophist: 'Of that, Socrates, you may be assured;-think only of the ambition of men, and you will wonder at the senselessness of their ways, unless you consider how they are stirred by the love of an immortality of fame. They are ready to run all risks greater far than they would have for their children, and to spend money and undergo any sort of toil, and even to die, for the sake of leaving behind them a name which shall be eternal. Do you imagine that Alcestis would have died to save Admetus, or Achilles to avenge Patroclus, or your own Codrus in order to preserve the kingdom for his sons, if they had not imagined that the memory of their virtues, which still survives among us, would be immortal? Nay,' she said, 'I am persuaded that all men do all things, and the better they are the more they do them, in hope of the glorious fame of immortal virtue; for they desire the immortal.

Those who are pregnant in the body only, betake themselves to women and beget children-this is the character of their love; their offspring, as they hope, will preserve their memory and giving them the blessedness and immortality which they desire in the future. But souls which are pregnant-for there certainly are men who are more creative in their souls than in their bodies conceive that which is proper for the soul to conceive or contain. And what are these conceptions? -wisdom and virtue in general. And such creators are poets and all artists who are deserving of the name inventor. But the greatest and fairest sort of wisdom by far is that which is concerned with the ordering of states and families, and which is called temperance and justice. And he who in youth has the seed of these implanted in him and is himself inspired, when he comes to maturity desires to beget and generate. He wanders about seeking beauty that he may beget offspring-for in deformity he will beget nothing-and naturally embraces the beautiful rather than the deformed body; above all when he finds fair and noble and well-nurtured soul, he embraces the two in one person, and to such an one he is full of speech about virtue and the nature and pursuits of a good man; and he tries to educate him; and at the touch of the beautiful which is ever present to his memory, even when absent, he brings forth that which he had conceived long before, and in company with him tends that which he brings forth; and they are married by a far nearer tie and have a closer friendship than those who beget mortal children, for the children who are their common offspring are fairer and more immortal. Who, when he thinks of Homer and Hesiod and other great poets, would not rather have their children than ordinary human ones? Who would not emulate them in the creation of children such as theirs, which have preserved their memory and given them everlasting glory? Or who would not have such children as Lycurgus left behind him to be the saviours, not only of Lacedaemon, but of Hellas, as one may say? There is Solon, too, who is the revered father of Athenian laws; and many others there are in many other places, both among hellenes and barbarians, who have given to the world many noble works, and have been the parents of *virtue of every kind*[15]; and many temples have been raised in their honour for the sake of children such as theirs; which were never raised in honour of any one, for the sake of his mortal children.

'These are the lesser mysteries of love, into which even you, Socrates, may enter; to the greater and more hidden ones which are the crown of these, and to which, if you pursue them in a right spirit,

---

[15] Relationship with the Material Cause: love is perceived as a virtue as Diotima explained it to Socrates.

they will lead, I know not whether you will be able to attain. But I will do my utmost to inform you, and do you follow if you can. For he who would proceed aright in this matter should begin in youth to visit beautiful forms; and first, if he be guided by his instructor aright, to love one such form only-out of that he should create fair thoughts; and soon he will of himself perceive that the beauty of one form is akin to the beauty of another; and then if beauty of form in general is his pursuit, how foolish would he be not to recognize that the beauty in every form is and the same! And when he perceives this he will abate his violent love of the one, which he will despise and deem a small thing, and will become *a lover of all beautiful forms*[16]; in the next stage he will consider that the beauty of the mind is more honourable than the beauty of the outward form. So that if a virtuous soul have but a little comeliness, he will be content to love and tend him, and will search out and bring to the birth *thoughts which may improve the young*[17], until he is compelled to contemplate and see the beauty of *institutions and laws*[18], and to understand that the beauty of them all is of one family, and that personal beauty is a trifle; and after laws and institutions he will go on to the sciences, that he may see their beauty, being not like a servant in love with the beauty of one youth or man or institution, himself a slave mean and narrow-minded, but drawing towards and contemplating the vast sea of beauty, he will create many fair and noble thoughts and notions in boundless love of wisdom; until on that shore he grows and waxes strong, *and at last the vision is revealed to him of a single science, which is the science*

---

[16] Relationship with the Formative Cause

[17] Relationship with the Efficient Cause

[18] Relationship with the Final Cause

of beauty everywhere[19]. To this I will proceed; please to give me your very best attention:

'He who has been instructed thus far in the things of love, and who has learned to see the beautiful in due order and succession, when he comes toward the end will suddenly perceive a nature of wondrous beauty (and this, Socrates, is the final cause of all our former toils)-a nature which in the first place is everlasting, not growing and decaying, or waxing and waning; secondly, not fair in one point of view and foul in another, or at one time or in one relation or at one place fair, at another time or in another relation or at another place foul, as if fair to some and-foul to others, or in the likeness of a face or hands or any other part of the bodily frame, or in any form of speech or knowledge, or existing in any other being, as for example, in an animal, or in heaven or in earth, or in any other place; but beauty absolute, separate, simple, and everlasting, which without diminution and without increase, or any change, is imparted to the ever-growing and perishing beauties of all other things. He who from these ascending under the influence of true love, begins to perceive that beauty, is not far from the end. And the true order of going, or being led by another, to the things of love, is to begin from the beauties of earth and mount upwards for the sake of that other beauty, using these as steps only, and from one going on to two, and from two to all fair forms, and from fair forms to fair practices, and from fair practices to fair notions, until from fair notions he arrives at the notion of absolute beauty, and at last knows what the essence of beauty is. This, my dear Socrates,' said the stranger of Mantineia, 'is that life above all others which man should live, in the contemplation of beauty

---

[19] Relationship with the Cause of Maturity

absolute; a beauty which if you once beheld, you would see not to be after the measure of gold, and garments, and fair boys and youths, whose presence now entrances you; and you and many a one would be content to live seeing them only and conversing with them without meat or drink, if that were possible-you only want to look at them and to be with them. But what if man had eyes to see the true beauty-the divine beauty, I mean, pure and dear and unalloyed, not clogged with the pollutions of mortality and all the colours and vanities of human life-thither looking, and holding converse with the true beauty simple and divine? Remember how in that communion only, beholding beauty with the eye of the mind, he will be enabled to bring forth, not images of beauty, but realities (for he has hold not of an image but of a reality), and bringing forth and nourishing true virtue to become the friend of God and be immortal, if mortal man may.

Would that be an ignoble life? [20]

---

[20] Plato, Symposium. Emphasis added.

III THE CAUSE OF MATURITY AND THE EXPLORATORY METHOD OF SCIENCE

> **SUMMARY OF THE MATURITY CAUSE:**
>
> *Question:* With which option? when addressing problems, difficulties, issues
>
> *Essence:* Mercy
>
> *Powers:* Intelligence, patience, discernment, compassion, empathy, volition
>
> *Situations* in which the human kingdom is being challenged to choose; but, in the case of the mineral, plant and animal kingdoms, they respond differentially to the stimuli received.
>
> **Decisions** being made at the individual and collective level: Governmental Institutions, Administration of private enterprises, NGOs, families or individuals
>
> *Kingdoms:* The decisions made by humans affecting the cycles in their own kingdom and (or) the other kingdoms
>
> *Stages and Phases:* The human decisions in relation to the stages of a plan or phases of a cycle
>
> *Sources of Wisdom:* What history or experience, the diverse disciplines of science and the world religions say about the issue to be addressed
>
> **Laws:** related to the inherent responsibilities of the assumed decision taking into account the effects in the other kingdoms
>
> *Relationships:* are those that interconnect the sources of wisdom
>
> *Parameters and indicators of progress:* Such as level of respect for the individual to exercise his (her) free will, level of commitment, number of pledges, and number of recommendations or referrals

With regard to the Maturity Cause and its relationship to the exploratory method, consider the example of the chair.

In order to understand *"Which options did the maker of the chair have?'*, one must think about: the options and the decisions made in each of the phases of the design process; the options and decisions made in selecting the materials for the chair; the choices and decisions made in each of the phases of the actual construction of the chair, including making it sturdy and safe.

What if, however, several chairs are broken and the question posed is to understand what happened so we do not make the same mistakes when building new ones? Then, we should look at all the decisions made during each of the phases of the cycle(s) that could be related to the specific damage of the chairs.

Let us then address the anxiety and fears related to our choices that arise from a mistaken identity. The essence of Mercy, as the cause of maturity, was the reason that help the author to bring into existence the set of notions and the corresponding set of human faculties associated with the exploratory and propositional methods of science. It will help us answer the question "Which options do I have once I reach the age of maturity?" by understanding the role of each of the components within the following philosophical argument.

Note: the author has added emphasis with *italics* to the words: *kingdom*, *journey*, *station* and *stage*; to help the reader connect key notions to themes that will be developed later on in this chapter.

PHILOSOPHICAL ARGUMENT

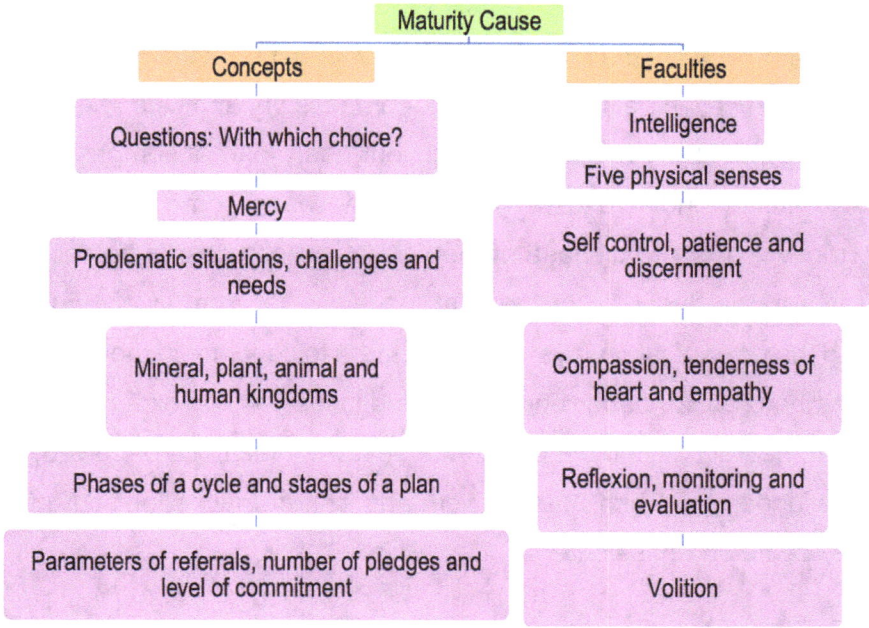

The Maturity Cause can be related to questions such as: *Which options do I have? Which is the best option? What is the strategy or master plan? Which decisions have to be made in order to address the situation? In which phase or stage did the problem originate?*

Mercy

The author has chosen mercy as the cause of governance and management because in the Kitáb-i-Aqdas "… He formally ordains the institution of the "House of Justice", defines its functions, fixes its revenues, and designates its members as the "Men of Justice", the "Deputies of God", the "Trustees of the All-Merciful""[21]

We begin our search by exploring the meaning of mercy at a more fundamental level. Mercy's synonyms are pity, compassion, kindness,

---

[21] Bahá'u'lláh, The Kitab-i-Aqdas, p. 13.

clemency, commiseration, humanitarianism, generosity, assistance, and indulgence.

> O ye beloved of the Lord! The Kingdom of God is founded upon equity and justice, and also upon mercy, compassion, and kindness to every living soul. Strive ye then with all your heart to treat compassionately all humankind -- except for those who have some selfish, private motive, or some disease of the soul. Kindness cannot be shown the tyrant, the deceiver, or the thief, because, far from awakening them to the error of their ways, it maketh them to continue in their perversity as before. No matter how much kindliness ye may expend upon the liar, he will but lie the more, for he believeth you to be deceived, while ye understand him but too well, and only remain silent out of your extreme compassion.[22]

Concerning the capacity of mercy to penetrate all of creation, Bahá'u'lláh says: "Look not upon the creatures of God except with the eye of kindliness and of mercy, for Our loving providence hath *pervaded* all created things, and *Our grace encompassed the earth and the heavens*."[23]

> All praise to the unity of God, and all honor to Him, the sovereign Lord, the incomparable and all-glorious Ruler of the universe, Who, out of utter nothingness, hath created the reality of all things, Who, from naught, hath brought into being the most refined and subtle elements of His creation, and Who, rescuing His creatures from the abasement of remoteness and the perils of ultimate extinction, hath received them into His *kingdom* of incorruptible glory. *Nothing short of His all-encompassing grace, His all-pervading mercy*, could have possibly achieved it. How could it, otherwise, have been possible for sheer

---

[22] Abdu'l-Baha, Selections from the Writings of Abdu'l-Baha, p. 158.

[23] Gleanings, p. 33. Emphasis Added.

nothingness to have acquired by itself the worthiness and capacity to emerge from its state of non-existence into the realm of being?[24]

By complementing the above quote with the following quotes, which are only a small sample of the significance of mercy, we can begin to grasp its importance when opting for alternatives:

"O son of man! If thine eyes be turned towards mercy, forsake the things that profit thee, and cleave unto that which will profit mankind. And if thine eyes be turned towards justice, choose thou for thy neighbor that which thou choosest for thyself."[25]

> Thus have We recounted unto you the tales of the one true God, and sent down unto you the things He had preordained, that haply ye may ask forgiveness of Him, may return unto Him, may truly repent, may realize your misdeeds, may shake off your slumber, may be roused from your heedlessness, may atone for the things that have escaped you, and be of them that do good. Let him who will, acknowledge the truth of My words; and as to him that willeth not, let him turn aside. My sole duty is to remind you of your failure in duty towards the Cause of God, if perchance ye may be of them that heed My warning.
>
> Wherefore, hearken ye unto My speech, and return ye to God and repent, that He, through His grace, may have mercy upon you, may wash away your sins, and forgive your trespasses. The greatness of His *mercy* surpasseth the fury of His wrath, and *His grace encompasseth all who have been called into being* and been clothed with the robe of life, be they of the past or of the future.[26]

---

[24] Bahá'u'lláh, Gleanings, pp. 64-65. Emphasis Added.

[25] Bahá'u'lláh, Epistle to the Son of the Wolf, pp. 29 -30.

[26] Bahá'u'lláh, Gleanings, p. 130. Emphasis Added.

Wisdom

We beseech the one true God, magnified be His glory, to enable us to recognize Him *Whose unerring wisdom pervadeth all things* and that we may acknowledge His truth. For once one hath recognized Him and borne witness to His Reality, one will no longer be troubled by the idle fancies and vain imaginings of men. The divine Physician hath the pulse of mankind within His almighty grasp. At one time He may well deem fit to sever certain infected limbs, that the disease may not spread to other parts of the body. This would be the very essence of mercy and compassion, and to none is given the right to object, for He is indeed the All-Knowing, the All-Seeing.[27]

Power

In reference to the concept of power, Bahá'u'lláh says:

Thou seest, O God of *mercy*, Thou *Whose power pervadeth all created things*, these servants of Thine, Thy thralls, who, according to the good-pleasure of Thy Will, observe in the daytime the fast prescribed by Thee, who arise, at the earliest dawn of day, to make mention of Thy Name, and to celebrate Thy praise, in the hope of obtaining their share of the goodly things that are treasured up within the treasuries of Thy grace and bounty.[28]

God, because of His mercy and His power, lets us know His teachings. 'Abdu'l-Bahá in an talk entitled *The Holy Spirit, the Intermediary Power between God and Man* says: "Man, then, is in extreme need of the only *Power* by which he is able to receive help from the Divine Reality, that *Power* alone bringing him into contact with the Source of all life."[29]

---

[27] Bahá'u'lláh, Tabernacle of Unity. P.44. Emphasis Added.

[28] Ibid. p. 299. Emphasis Added.

[29] Paris Talks, pp. 724-25. Emphasis Added.

The Divine Reality is Unthinkable, Limitless, Eternal, Immortal and Invisible.

The world of creation is bound by natural law, finite and mortal.

The Infinite Reality cannot be said to ascend or descend. It is beyond the understanding of man and cannot be described in terms which apply to the phenomenal sphere of the created world.

Man, then, is in extreme need of the only Power by which he is able to receive help from the Divine Reality, that Power alone bringing him into contact with the Source of all life.

An intermediary is needed to bring two extremes into relation with each other. Riches and poverty, plenty and need: without an intermediary power there could be no relation between these pairs of opposites.

So we can say there must be a Mediator between God and Man, and this is none other than the Holy Spirit, which brings the created earth into relation with the 'Unthinkable One', the Divine Reality.

The Divine Reality may be likened to the sun and the Holy Spirit to the rays of the sun. As the rays of the sun bring the light and warmth of the sun to the earth, giving life to all created beings, so do the 'Manifestations' bring the power of the Holy Spirit from the Divine Sun of Reality to give light and life to the souls of men.

Behold, there is an intermediary necessary between the sun and the earth; the sun does not descend to the earth, neither does the earth ascend to the sun. This contact is made by the rays of the sun which bring light and warmth and heat.

The Holy Spirit is the Light from the Sun of Truth bringing, by its infinite power, life and illumination to all mankind, flooding all souls with Divine Radiance, conveying the blessings of God's mercy to the whole world. ...

Likewise the Holy Spirit is the very cause of the life of man; without the Holy Spirit he would have no intellect, he would be unable to acquire his scientific knowledge by which his great influence over the rest of creation is gained. The illumination of the Holy Spirit gives to man the power of thought, and enables him to make discoveries by which he bends the laws of nature to his will.

The Holy Spirit it is which, through the mediation of the Prophets of God, teaches spiritual virtues to man and enables him to attain Eternal Life.[30]

In order to choose the most powerful strategy, we must examine the most relevant findings and make an effort to word them with a convincing argument. Without any doubt, Bahá'í institutions exercise authority, but the power to implement authority depends highly on the level of comprehension, understanding and willingness of each individual, whether or not a member of the institution, to act on what has been agreed. All of us are part of the Bahá'í administrative order in the sense that when the meeting of the institution dissolves, all are but mere citizens who abide by the decisions of the institutions.

Mercy is a key feature of the representatives of the Bahá'í communities, the members of the Spiritual Assemblies, because they are the trustees of the Merciful. The exercise of authority of the Bahá'í Administrative Order is not based on coercion or punishment. Everyone should help others understand the wisdom behind the decisions made by Local Spiritual Assemblies. The individual should show mercy and help the institutions to educate humanity so that they can have a comprehensive foundation to discern. This should be done with the outmost respect of the free will of the individual. Individuals are empowered when decide to act

---

[30] ibid. pp. 724-25.

using their own volition! Each person has to decide and assume the consequences of his choices. If a Bahá'í persists in wrongdoing, causing damages to others, it is the role of the Bahá'í institutions to exercise justice.

Scientists have the option to dedicate their lives to exploit people or to establish economic justice, to create even more powerful instruments of human destruction or to dedicate their creativity to peaceful conflict resolution.

Situations with Challenges, Problems, Tests and Difficulties

"There is no philosophical high-road in science, with epistemological signposts. No, we are in a jungle and find our way by trial and error, building our roads behind us as we proceed. We do not find sign-posts at cross-roads, but our own scouts erect them, to help the rest."[31]

"The stumbling way in which even the ablest of the scientists in every generation have had to fight through thickets of erroneous observations, misleading generalizations, inadequate formulations, and unconscious prejudices rarely appreciated by those who obtain their scientific knowledge from textbooks."[32]

Bahá'u'lláh says:

> The All-Knowing Physician hath His finger on the pulse of mankind. He perceiveth the *disease*, and prescribeth, in His unerring *wisdom*, the remedy. Every age hath its own *problem*, and every soul its particular aspiration. The remedy the world needeth in its present-day *afflictions* can never be the same as that which a subsequent age may require. Be anxiously concerned with the *needs* of the age ye live in, and centre your deliberations on its exigencies and requirements.[33]

---

[31] Born, Max, Thematic Origins of Scientific Thought, p.7.

[32] Conant, James Bryant.

[33] Bahá'u'lláh, Tabernacle of Unity, p.25. Emphasis Added.

Challenges are blessings:

The one true God well knoweth, and all the company of His trusted ones testify, that this Wronged One hath, at all times, been faced with dire peril. But for the *tribulations* that have touched Me in the path of God, life would have held no sweetness for Me, and My existence would have profited Me nothing. For them who are endued with discernment, and whose eyes are fixed upon the Sublime Vision, it is no secret that I have been, most of the days of My life, even as a slave, sitting under a sword hanging on a thread, knowing not whether it would fall soon or late upon him. And yet, notwithstanding all this We render thanks unto God, the Lord of the worlds. Mine inner tongue reciteth, in the daytime and in the night-season, this prayer: 'Glory to Thee, O my God! But for the *tribulations* which are sustained in Thy path, how could Thy true lovers be recognized; and were it not for the *trials* which are borne for love of Thee, how could the *station* of such as yearn for Thee be revealed?'[34]

What could be the meaning of words like development, progress, cleanliness, faith, and peace without issues or problems? What could be the meaning of words like, science, religion, or a profession in the absence of needs to be satisfied? We all, individually and collectively, must excel in the solution of every problem that seems to be important. By excelling in confronting any important problem, we are presented with a unique and unparalleled opportunity for growth and resolution; an opportunity for change, that will affect our neighborhoods, our countries and the world for generations to come. We become historical constructors of new civilization.

Sources of Wisdom

The sources of wisdom are: what history, experience, world religions and the diverse disciplines of science say about the issue to be solved. In

---

[34] Bahá'u'lláh, Epistle to the Son of the Wolf, p. 94. Emphasis Added.

order to choose the best strategy to address a problem it is advisable to transcend the narrow search within one discipline.

Intelligence

Human beings are endowed with intelligence to make wise decisions. 'Abdu'l-Bahá, the Master, said in Paris in the year 1911:

> God's greatest gift to man is that of intellect, or understanding.
>
> The understanding is the power by which man acquires his knowledge of the several *kingdoms* of creation, and of various *stages* of existence, as well as of much which is invisible.
>
> Possessing this gift, he is, in himself, the sum of earlier creations -- he is able to get into touch with those *kingdoms*; and by this gift, he can frequently, through his scientific knowledge, reach out with prophetic vision.
>
> Intellect is, in truth, the most precious gift bestowed upon man by the Divine Bounty. Man alone, among created beings, has this wonderful power.
>
> All creation, preceding Man, is bound by the stern law of nature. ... [35]

"Like the animal, man possesses the faculties of the senses, is subject to heat, cold, hunger, thirst, etc.; unlike the animal, man has a rational soul, the human intelligence."[36]

"The animal may develop a wonderful degree of intelligence, but it can never attain the powers of ideation and conscious reflection which belong to man."[37]

---

[35] Paris Talks, p. 715. Emphasis Added.

[36] ibid. p. 746.

[37] 'Abdu'l-Bahá, The Promulgation of Universal Peace, p. 954.

Human Kingdom and other Kingdoms in Nature

The kingdoms of nature and those in authority, in any administrative order, are also important in the Maturity Cause. The decisions of those in command, when we want to explore a problematic situation, should also be considered in terms of how the other kingdoms are affected. Of course, an individual's decision also should be taken into account.

> "All created things have their degree or *stage* of maturity. The period of maturity in the life of a tree is the time of its fruit-bearing. The maturity of a plant is the time of its blossoming and flower. The animal attains a *stage* of full growth and completeness, and in the human *kingdom* man reaches his maturity when the lights of intelligence have their greatest power and development."[38]

Also the difference of conditions in the world of beings is an obstacle to comprehension. For example, this mineral belongs to the mineral kingdom; however far it may rise, it can never comprehend the power of growth. The plants, the trees, whatever progress they may make, cannot conceive of the power of sight or the powers of the other senses; and the animal cannot imagine the condition of man—that is to say, his spiritual powers. Difference of condition is an obstacle to knowledge; the inferior degree cannot comprehend the superior degree. How then can the phenomenal reality comprehend the Preexistent Reality? Knowing God, therefore, means the comprehension and the knowledge of His attributes, and not of His Reality. This knowledge of the attributes is also proportioned to the capacity and power of man; it is not absolute. Philosophy consists in comprehending the reality of things as they exist, according to the capacity and the power of man. For the phenomenal reality can comprehend the Preexistent attributes only to the extent of the human capacity. The mystery of Divinity is sanctified and purified from the

---

[38] 'Abdu'l-Bahá, Foundations of World Unit, p. 9. Emphasis Added.

comprehension of the beings, for all that comes to the imagination is that which man understands, and the power of the understanding of man does not embrace the Reality of the Divine Essence. All that man is able to understand are the attributes of Divinity, the radiance of which appears and is visible in the world and within men's souls.[39]

Then, especially in a globalized, interconnected world, we should be aware of how our options may affect the other kingdoms.

Stages and Stations

In *The Valley of True Poverty and Absolute Nothingness*, which describes the last phase of human earthly progress, Bahá'u'lláh says:

> This *station* is the dying from self and the living in God, the being poor in self and rich in the Desired One. Poverty as here referred to signifieth being poor in the things of the created world, rich in the things of God's world. For when the true lover and devoted friend reacheth to the presence of the Beloved, the sparkling beauty of the Loved One and the fire of the lover's heart will kindle a blaze and burn away all veils and wrappings. Yea, all he hath, from heart to skin, will be set aflame, so that nothing will remain save the Friend.
>
> When the qualities of the Ancient of Days stood revealed, Then the qualities of earthly things did Moses burn away.[40]
>
> In this Valley, the wayfarer leaveth behind him the *stages* of the 'oneness of Being and Manifestation' and reacheth a oneness that is sanctified above these two *stations*. Ecstasy alone can encompass this theme, not utterance nor argument; and whosoever hath dwelt at

---

[39] 'Abdu'l-Bahá, Some Answered Questions, pp. 221-22.

[40] The Seven Valleys and The Four Valleys, p. 36. Emphasis Added.

this *stage* of the *journey*, or caught a breath from this garden land, knoweth whereof We speak.

In all these *journeys* the traveler must stray not the breadth of a hair from the 'Law,' for this is indeed the secret of the 'Path' and the fruit of the Tree of 'Truth'; and in all these *stages* he must cling to the robe of obedience to the commandments, and hold fast to the cord of shunning all forbidden things, that he may be nourished from the cup of the Law and informed of the mysteries of Truth.[41]

The Maturity Cause is necessary for the preparation for adulthood and life after death. As we mature we gain more understanding and discernment to explore different alternatives to solve problems. The most desirable solution is that which compromises our will to the highest degree, in order to reach what God has ordained for the people on earth. A point is reached when the individual places his absolute trust in God's teachings because he has found immense wisdom every time he has explored any one of them.

In the aforementioned quotes I have *added emphasis* to the words *journey*, *station* and *stage*, which suggest phases within a journey or regularity within a cycle. It is clear that all problems, difficulties and needs could be connected to one or more cycles or journeys. The phase of the cycle or the stage of the journey to be explored will be the one considered critical in relation to the problem to be solved, because the emerging characteristics can be traced to the problem's features.

When we study the sources of wisdom in order to address a certain situation we should carefully examine the effect of the decisions made by the human kingdom in the other kingdoms, in each one of the stages or phases. Taking into account the developmental stages of a situation, of an illness, of an animal, of a plant or of a human being, are key to making the right choice. The same idea applies to the stations within a journey.

---

[41] ibid. p. 39. Emphasis Added.

Relationships

In Wikipedia we find a fascinating reflection: In "the 1985 CBC series "*A Planet for the Taking*", Dr. David Suzuki explored the Old Testament roots of anthropocentrism and how it shaped our view of non-human animals. Some Christian proponents of anthropocentrism base their belief on the Bible, such as the verse 1:26 in the Book of Genesis:

> And God said, Let us make man in our image, after our likeness: and let them have dominion over the fish of the sea, and over the fowl of the air, and over the cattle, and over all the earth, and over every creeping thing that creepeth upon the earth.
>
> The use of the word 'dominion' in the *Genesis* is controversial. Many Biblical scholars, especially Roman Catholic and other non-Protestant Christians, consider this to be a flawed translation of a word meaning 'stewardship', which would indicate that mankind should take care of the earth and its various forms of life, but is not inherently better than any other form of life. The current Latin Vulgate, the official Bible of the Catholic Christian church, states that God holds man responsible for the care and fate of all earthly creatures.[42]

We can perceive other relationships of governance in the manner with which power is exercised: those who dominate and impose their will by force, versus those who empower the masses to accept what has been decided by evaluating the wisdom of the decisions taken by those in authority.

Other relationships to be considered in the Maturity Cause are those that interconnect the sources of wisdom.

---

[42] "anthropocentrism." Wikipedia.

Laws

In this fifth cause the laws are related to the inherent responsibilities of the assumed decision, those that apply to a being when it passes from one stage of existence to the next and those that apply when it reaches the phase of extinction.

Parameters and Indicators of Progress

The follow up controls of parameters and indicators and the mechanisms of regulation and evaluation of performance, in the phases within the cycle or at the stages of a journey, are crucial to exploring the situation's evolution and proposing alternative solutions.

It is important to say, that in *En Síntesis la Ciencia es Amor* (1992), I included parameters and indicators for each one of the four causes (Final, Material, Efficient and Formative) taught by the Greek Philosophers, not because I was conscious of the Cause of Maturity, but because of interest as a scientist. Later on, while I was working on the Cause of Maturity (1999), I realize, obviously, that parameters and indicators where very important to choose among alternatives as pivotal for Governance and Management.

	THE CAUSE OF MATURITY AND THE EXPLORATORY AND PROPOSITIONAL METHODS OF SCIENCE	
1	**Question:** With which option? when addressing problems, difficulties, issues and needs	CAUSE OF MATURITY
2	**Essence**: Mercy. **Sources of Wisdom**: What history or experience, the diverse disciplines of science and the world religions say about the issue to be addressed **Kingdoms, Cycles and phases**	MATERIAL CAUSE

3	**Laws:** related to the inherent responsibilities of the assumed decision taking into account the effects in the other Kingdoms and in the Human Kingdom. **Goal**: Stewardship in the management of the trust	FINAL CAUSE
4	**Powers**: Intelligence, patience, self-control, discernment, compassion, empathy, volition **Decisions made at the individual and collective level:** Governmental Institutions, Administration of private enterprises, NGOs, families or individuals. The human decisions in relation to the stages of a plan or phases of a cycle **The spontaneous differential response** to the stimuli received by the different kingdoms **The effects of the decisions made by humans in all kingdoms** **The developmental stage** of the individual, plant or animal **Monitoring and evaluating** the strategy and its stages and the cycle and its phases	EFFICIENT CAUSE
5	**Situations** in which the human kingdom is being challenged to choose or, in the case of the mineral, plant and animal kingdoms, the spontaneous differential response to the stimuli received. **Relationships:** are those that interconnect the sources of wisdom	FORMATIVE CAUSE
6	**Parameters and indicators of progress**: Such as level of respect for the individual to exercise his (her) free will, level of commitment, number of pledges, and number of recommendations or referrals	CAUSE OF MATURITY

Applying the Single Science to the Cause of Maturity

The table below shows the methodology followed in the construction of the "New Approach to Science" by inter-weaving the causes horizontally and vertically. In order for the reader to understand the reason of the placement of the following acceptations in one of the Rows (1 to 6), it is important to comprehend how the Cause of Maturity is interwoven with the other causes.

Having the basis of a conceptual framework for the Cause of Maturity and learned in the single science proposed above in "Classes and Categories as foundation of Philosophy", that we are supposed to classify each notion of interest for the solution of the problem in one of the causes; the following sample of a general accepted meaning of a word could be placed in the Cause of Maturity:

Membrane: "A cell membrane surrounds and protects the contents of a cell. It controls which substances can enter and exit the cell. The membrane also gives a cell its shape and enables the cell to attach to other cells, forming tissues."[43] (In row 4 the intersection of The Cause of Maturity and The Efficient Cause will be the right place because to *control* is an action, but, could also be placed in row 3 the intersection of The Cause of Maturity and The Final Cause because the membrane surrounds and *protects* the contents of a cell, but, could also be placed in row 5 the intersection of The Cause of Maturity and The Formative Cause because the membrane also gives a cell its *shape* and enables the cell to *attach* to other cells, *forming* tissues)

Plasma membrane: "The limiting surface of the cytoplasm of a eukaryotic cell. It consists of a phospholipid bilayer with a variety of embedded molecules that act as channels and pumps, selectively moving particular molecules into and out of the cell.

---

[43] "Membrane." <https://quizlet.com>.

Surface molecules on the plasma membrane allow specific recognition of each particular cell type. Phospholipids have a head group, which is attracted to water, and a tail group, which is made up of a long hydrocarbon chain repelled by water. Phospholipids are the primary constituent of the lipid bilayers of cells."[44] (In row 2 the intersection of The Cause of Maturity and The Material Cause if looking into the composition of the plasma membrane consisting of a *phospholipid* bilayer with a variety of embedded *molecules*, but, could also be placed in row 4 the intersection of The Cause of Maturity and The Efficient Cause if we focus our attention into the "embedded molecules *that act* as channels and pumps, which function is of selectively *moving* particular molecules into and out of the cell but, could also be placed in row 4 the intersection of The Cause of Maturity and The Cause of Maturity because its function is of *selectively* moving particular molecules into and out of the cell)

Molt: "To cast off the outer covering. Birds molt old feathers once or twice a year. Reptiles molt old skin, and arthropods cast off the entire cuticle. Mammals also molt hair, but the term shed is usually used in this case."[45] (In row 4 the intersection of The Cause of Maturity and The Efficient Cause because molting is a process in which feathers, skin and hair are the leftover)

Carbon cycle: "A cycle composed of two primary processes, photosynthesis and respiration. Photosynthesis produces oxygen and glucose from carbon dioxide and water. Respiration reverses this by creating carbon dioxide and water from glucose and oxygen."[46] (In row 2 the intersection of The Cause of Maturity and The Material Cause because it is *composed* of two primary

---

[44] "plasma membrane." Eugene M. Mccarthy, Online Biology Dictionary.

[45] "molt." Eugene M. Mccarthy, Online Biology Dictionary.

[46] "carbon cycle." Eugene M. Mccarthy, Online Biology Dictionary.

processes, but, could also be placed in row 4 the intersection of The Cause of Maturity and The Efficient Cause if we are looking at the processes and their results)

Lytic cycle: "The lytic cycle results in the destruction of the infected cell and its membrane."[47] (In row 2 the intersection of The Cause of Maturity and The Material Cause because that infected cell becomes part of the mineral kingdom, but, could also be placed in row 3 the intersection of The Cause of Maturity and The Final Cause because the destruction of the infected cell is a consequence of the action of the immune system, but, could also be placed in row 1 the intersection of The Cause of Maturity and "The Cause of Maturity" because the problem is that the cell is infected)

Pupae: "An intermediate usually quiescent stage of a metamorphic insect (as a bee, moth, or beetle) that occurs between the larva and the imago, is usually enclosed in a cocoon or protective covering, and undergoes internal changes by which larval structures are replaced by those typical of the imago."[48] (In row 4 the intersection of The Cause of Maturity and The Efficient Cause because it is a *developmental* stage undergoing internal *changes*, but, could also be placed in row 2 the intersection of The Cause of Maturity and The Material Cause because pupae is stage of the metamorphic cycle; in other words a cycle is composed of stages or phases)

The same grouping criteria applies to the below mentioned faculties associated with the Cause of Maturity.

---

[47] "lytic cycle." Wikipedia.

[48] "pupae." <https://quizlet.com>.

OTHER HUMAN FACULTIES LINKED TO THE MATURITY CAUSE

Besides the power of intelligence, mercy, compassion, tenderness of heart, empathy and power, human beings have also been endowed with:

Senses

"The outward powers are five: the power of sight, of hearing, of taste, of smell, and of touch."[49] With our outward senses we perceive the stages of the journey and the phases of the cycle.

Discernment and Self-control

The Maturity Cause is linked to the question: *With which option?* In order to enable us to make thoughtful choices, we have been endowed with discernment, patience and self-control.

When we are discerning, we should carefully examine how we are going to affect the human kingdom and the other kingdoms in each one of the stages or phases.

Because mercy is inherent to all beings in creation, it is also important to acknowledge that, while in the human kingdom there is discernment in the decision making process, the mineral, vegetable and animal kingdoms respond differentially in accordance with the stimulus received from the environment's cycles, such as: deciduous trees lose their leaves in the fall and grow new ones in the spring; bears' spontaneous differential response during winter is hibernation; also the *differential response* of *minerals* to heating and cooling: water and stones become warmer during sunny days.

Another way to perceive a differential response is thru homeostasis, which is the property of a cell or a whole organism to seek and maintain a condition of balance or equilibrium within its internal environment, even when faced with external changes. Homeostasis is a "Feedback Cycle ... in which a variable is regulated and the level of the variable impacts the

---

[49] 'Abdu'l-Bahá, Some Answered Questions, p. 56.

direction in which the variable changes (i.e. increases or decreases), even if there is not clearly identified loop components."[50]

> Homeostasis is the property of a system within the body of an organism in which a variable, such as the concentration of a substance in solution, is actively regulated to remain very nearly constant. Examples of homeostasis include the regulation of the body temperature of an animal, the pH of its extracellular fluids, or the concentrations of sodium (Na^+), potassium (K^+) and calcium (Ca^{2+}) ions as well as that of glucose in the blood plasma, despite changes in the animal's environment, or what it has eaten, or what it is doing (for example, resting or exercising).[51]

This differential response within the mineral, vegetable and animal kingdoms takes place without the exercise of free will. Human beings have been endowed with a conscious choosing process after considering different alternatives, implying self-control and volition. With these faculties, we humans are able to alter the differential response of the other kingdoms, for example when we plant a forest or when we build a dam in a river.

But not all of our relation with the other kingdoms requires discernment and free will. When an insect bites us our immune system spontaneously responds differentially in accordance with the chemical compounds and virus or bacteria introduced into our system. When we breathe (Human Kingdom) we inhale oxygen (Material Kingdom), previously exhaled by the Plant Kingdom during photosynthesis, and each living cell of our organism (Human Kingdom) responds differentially when oxygen is present in the blood stream, and then we exhale carbon dioxide which plants take in during photosynthesis. Of course, we can consciously alter our breathing when we decide to run.

---

[50] https://courses.lumenlearning.com/ap1/chapter/homeostasis-and-feedback-loops/.

[51] Wikipedia, Homeostasis.

Our reflexes are also an example of spontaneous differential response to the stimuli received from the environment without the exercise of discernment and volition.

Bahá'u'lláh expresses:

> Know thou that, according to what thy Lord, the Lord of all men, hath decreed in His Book, the favors vouchsafed by Him unto mankind have been, and will ever remain, limitless in their range. First and foremost among these favors, which the Almighty hath conferred upon man, is the gift of understanding. His purpose in conferring such a gift is none other except to enable His creature to know and recognize the one true God -- exalted be His glory. *This gift giveth man the power to discern the truth in all things*, leadeth him to that which is right, and helpeth him to discover the secrets of creation.[52]

Abdu'l-Bahá, the Master, mentions self-control in the following quote:

> Make ye then a mighty effort, that the purity and sanctity which, above all else, are cherished by 'Abdu'l-Bahá, shall distinguish the people of Bahá; that in every kind of excellence the people of God shall surpass all other human beings; that both outwardly and inwardly they shall prove superior to the rest; that for purity, immaculacy, refinement, and the preservation of health, they shall be leaders in the vanguard of those who know. And that by their freedom from enslavement, their knowledge, *their self-control*, they shall be first among the pure, the free and the wise.[53]

Tenderness, Compassion and Empathy

Finally, to be wiser, men should subordinate their decisions to the following points of reference:

---

[52] Gleanings, p. 193. Emphasis Added.

[53] 'Abdu'l-Bahá, Selections, p. 150. Emphasis Added.

As regards the constitution of the House of Justice, Bahá'u'lláh addresses the men. He says: 'O ye men of the House of Justice!'

But when its members are to be elected, the right which belongs to women, so far as their voting and their voice is concerned, is indisputable. When the women attain to the ultimate degree of progress, then, according to the exigency of the time and place and their great capacity, they shall obtain extraordinary privileges. Be ye confident on these accounts. His Holiness Bahá'u'lláh has greatly strengthened the cause of women, and the rights and privileges of women is one of the greatest principles of 'Abdu'l-Bahá. Rest ye assured! Ere long the days shall come when the men addressing the women, shall say: 'Blessed are ye! Blessed are ye! Verily ye are worthy of every gift. Verily ye deserve to adorn your heads with the crown of everlasting glory, because in sciences and arts, in virtues and perfections ye shall become equal to man, and as regards *tenderness of heart and the abundance of mercy and sympathy* ye are superior'.[54]

My suggestion is: the abundance of mercy, tenderness and sympathy demonstrated by women is the reason why they should be welcome to participate in greater number, until they reach majority, in the administrative level of all human affairs even at the international level, except in the Universal House of Justice as ordained by Bahá'u'lláh. To clarify the reason of such exception, Shoghi Effendi, in a letter written on his behalf to an individual believer (28 July 1936), provided the following authoritative elaboration of this theme:

As regards your question concerning the membership of the Universal House of Justice: there is a Tablet from 'Abdul-Bahá in which He definitely states that the membership of the Universal House is confined to men, and that the wisdom of it will be fully

---

[54] 'Abdu'l-Bahá, Paris Talks, p. 794. Emphasis Added.

revealed and appreciated in the future. In the local as well as the national Houses of Justice, however, women have the full right of membership. It is, therefore, only to the International House that they cannot be elected. The Bahá'ís should accept this statement of the Master in a spirit of deep faith, confident that there is a divine guidance and wisdom behind it which will be gradually unfolded to the eyes of the world.[55]

Rather than expressing mortal ideas about mercy and the exercise of authority, in relation to tenderness and compassion, let us continue reflecting upon the Word of Bahá'u'lláh Himself:

> I swear by the beauty of the Well-Beloved! This is the mercy that hath encompassed the entire creation, the Day whereon the grace of God hath permeated and pervaded all things. The living waters of My mercy, O Ali, are fast pouring down, and Mine heart is melting with the heat of My tenderness and love. At no time have I been able to reconcile Myself to the afflictions befalling My loved ones, or to any trouble that could becloud the joy of their hearts.
>
> Every time My name 'the All-Merciful' was told that one of My lovers had breathed a word that runneth counter to My wish, it repaired, grief-stricken and disconsolate to its abode; and whenever My name 'the Concealer' discovered that one of My followers had inflicted any shame or humiliation on his neighbor, it, likewise, turned back chagrined and sorrowful to its retreats of glory, and there wept and mourned with a sore lamentation. And whenever My name 'the Ever-Forgiving' perceived that any one of My friends had committed any transgression, it cried out in its great distress, and, overcome with

---

[55] qtd. in Universal House of Justice, A Compilation on Women.

anguish, fell upon the dust, and was borne away by a company of the invisible angels to its habitation in the realms above.[56]

God shows us the immensity of His mercy and it is our obligation to respond to His laws. Although He is the Merciful and the Concealer, we could be inclined to disobey; because man cannot stop the effects of His laws, we should opt for what He has ordained. The same applies when having to choose between what is just and what is not; between what is conducive to wealth and what is conducive to poverty; between science that teaches to discern and the one that teaches how to divide, exploit and oppress people. In the decision-making process, we should examine the options with the lamp of divine mercy, grace and kindness.

Two Natures

Although the virtues linked to the Maturity Cause have been explained above, there are other factors to consider when asking ourselves questions like: Which options do I have? Which is the best option? What is the strategy? Which decisions have to be made in order to address the situation? In which phase or stage did the problem originate?

It is of great transcendence to acknowledge the existence of two natures, a higher and a lower one, as an important element into the decision-making process.

"God has created all in His image and likeness. Shall we manifest hatred for His creatures and servants? This would be contrary to the will of God and according to the will of Satan, by which we mean the natural inclinations of the lower nature. This lower nature in man is symbolized as Satan -- the evil ego within us, not an evil personality outside."[57]

In man there are two natures; his spiritual or higher nature and his material or lower nature. In one he approaches God, in the other he lives for the world alone. Signs of both these natures are to be found

---

[56] Gleanings, pp. 308-09. Emphasis Added.

[57] 'Abdu'l-Bahá, The Promulgation of Universal Peace, p. 286.

in men. In his material aspect he expresses untruth, cruelty and injustice; all these are the outcome of his lower nature. The attributes of his Divine nature are shown forth in love, mercy, kindness, truth and justice, one and all being expressions of his higher nature. Every good habit, every noble quality belongs to man's spiritual nature, whereas all his imperfections and sinful actions are born of his material nature. If a man's Divine nature dominates his human nature, we have a saint.[58]

When you wish to reflect upon or consider a matter, you consult something within you. You say, shall I do it, or shall I not do it? Is it better to make this *journey* or abandon it? Whom do you consult? Who is within you deciding this question? Surely there is a distinct power, an *intelligent ego*. Were it not distinct from your *ego*, you would not be consulting it. It is greater than the faculty of thought. It is your spirit which teaches you, which advises and decides upon matters.[59]

The Heart's Role in the Decision Making Process Must also be Considered

'Every subject presented to a thoughtful audience must be supported by rational proofs and logical arguments. Proofs are of four kinds: first, through sense-perception; second, through the reasoning faculty; third, from traditional or scriptural authority; fourth, through the medium of inspiration. That is to say, there are four criterions or standards of judgment by which the human mind reaches its conclusions.[60]

'Abdu'l-Bahá then explains each of those criteria and demonstrates why they are individually unreliable, saying:

---

[58] 'Abdu'l-Bahá, Paris Talks, p. 18.

[59] 'Abdu'l-Bahá, The Promulgation of Universal Peace, p. 242. Emphasis Added.

[60] Foundations of World Unity, p. 86. Emphasis Added.

Consequently it has become evident that the four criteria or standards of judgment by which the human mind reaches its conclusions are faulty and inaccurate. All of them are liable to mistake and error in conclusions. But a statement presented to the mind accompanied by proofs which the senses can perceive to be correct, which the faculty of reason can accept, which is in accord with traditional authority *and sanctioned by the promptings of the heart*, can be adjudged and relied upon as perfectly correct, for it has been proved and tested by all the standards of judgment and found to be complete. When we apply but one test there are possibilities of mistake. This is self-evident and manifest.[61]

Free Will

After consulting our lower and higher natures in our hearts, we commit our free will, also called volition, to the solution that seems to be wise:

> And now, concerning thy question regarding the creation of man. Know thou that all men have been created in the nature made by God, the Guardian, the Self-Subsisting. Unto each one hath been prescribed a pre-ordained measure, as decreed in God's mighty and guarded Tablets. All that which ye potentially possess can, however, be manifested only as a result of your *own volition*.[62]

> Man alone has *freedom*, and, by his understanding or intellect, has been able to gain control of and adapt some of those natural laws to his own needs. By the power of his intellect he has discovered means by which he not only traverses great continents in express trains and crosses vast oceans in ships, but, like the fish he travels under water

---

[61] ibid. p.87.

[62] Bahá'u'lláh, Gleanings, p. 149. Emphasis Added.

in submarines, and, imitating the birds, he flies through the air in airships.[63]

"All of us know that international peace is good, that it is conducive to human welfare and the glory of man but *volition* and action are necessary before it can be established."[64]

The attainment of any object is conditioned upon knowledge, *volition* and action. Unless these three conditions are forthcoming there is no execution or accomplishment. In the erection of a house it is first necessary to know the ground and design the house suitable for it; second, to obtain the means or funds necessary for the construction; third, to actually build it.[65]

Sense of our Own Powerlessness when we Realize the Unfathomable Depths of His Wisdom in the Books of Revelation and Creation

Inspire them, O my Lord, with a sense of their *own powerlessness* before Him Who is the Manifestation of Thy Self, and teach them to recognize the poverty of their own nature in the face of the manifold tokens of Thy self-sufficiency and riches, that they may gather together round Thy Cause, and cling to the hem of Thy mercy, and cleave to the cord of the good-pleasure of Thy will.[66]

*

In conclusion, one of the phases of the cycle of scientific research is the exploratory phase; and the philosophical argument and the set of human faculties articulated to the Maturity Cause are ideal to be able to

---

[63] 'Abdu'l-Bahá, Paris Talks, p. 716. Emphasis Added.

[64] 'Abdu'l-Bahá, Foundations of World Unity, p. 26. Emphasis Added.

[65] ibid. p. 100. Emphasis Added.

[66] Bahá'u'lláh, Prayers and Meditations by Bahá'u'lláh, p. 47. Emphasis Added.

answer the question: Which options do I have once I reach the age of maturity?

IV SUMMARY OF THE FOUR CAUSES TAUGHT BY THE GREEK PHILOSOPHERS

Let us proceed now to study the four causes taught by the Greek philosophers; and explain in a comprehensive way why the notions mentioned in the summary below have been attached to each cause.

FINAL CAUSE:
- Why? For what?
- *Essence*: the power of law itself, i.e., that which is a condition to a certain aspect of life or nature, and explains its purpose, its mission
- Situations of risk and danger
- Pacts, agreements and covenants
- Natural laws
- Religious laws and ordinances
- Social order, laws and norms
- Advice from those with experience
- Individual and collective commitment to ideals
- Relationships of gratitude, loyalty, reciprocity, mutuality, of fear and protection; and also guilt, repentance, punishment and reward
- Parameters and indicators of protection, equality, prevention and security

EFFICIENT CAUSE:
- With what being? With whom? With what type?
- *Essence*: the Word of God
- Categories: genres, clusters, groups
- Animated and inanimated beings
- Forces: power of mind, energy, capacities, potential, talents, vocations, arguments, concepts
- To do: movements: methods, processes, activities, abilities, skills, arts, technologies, mechanisms
- Interacting entities within a system or a subsystem
- Fruits of labor: services, products and leftover materials or efforts
- Laws of thermodynamics, labor law and tools regulations. Grammar rules
- Relations of: cause and effect, logic and reason
- Parameters and indicators of efficiency and productivity

MATERIAL CAUSE:
- With what?
- *Essence*: the Spirit
- Specific properties of matter, peculiarities of things and characteristics
- Moral values, also called spiritual values
- Knowledge, beliefs and social values
- Elements, substances and raw materials
- Spiritual laws, the standards that regulate quality and laws of possession
- Specific lines of action or policies to consolidate ethics, knowledge, quality and efficacy
- Relationships of belonging, possession, distinction and characterization
- Parameters and indicators of efficacy and quality

FORMATIVE CAUSE:
- How is the arrangement?
- *Essence*: love, feelings and soul as its power
- General properties of matter: form, size, mass, temperature, movement and position in time and space
- Scientific principles related to organization, arrangement and design
- Principles of Unity in all Religions
- The 11 Principles of Unity in accordance with the teachings of Bahá'u'lláh[67]
- Laws related to unification and gravity
- Aspirations, desires and emotions
- Parameters and indicators of unity, beauty, symmetry, harmony, reciprocity and justice

---

[67] The Eleven Principles out of the Teaching of Bahá'u'lláh, explained by 'Abdu'l-Bahá in Paris: "The Search after Truth. The Unity of Mankind. Religion ought to be the Cause of Love and Affection. The Unity of Religion and Science. Abolition of Prejudices. Eliminating Extremes of Poverty and Wealth. Equality of Men before

the Law. Universal Peace. Non-Interference of Religion and Politics. Equality of Sex -- Education of Women. The Power of the Holy Spirit" (Paris Talks 123)

V THE FORMATIVE CAUSE AND THE FORMATIVE METHOD OF SCIENCE

In order to introduce the 'Formative Cause' let us continue with the example of the chair. To understand *How was the chair made?* it is necessary to consider, among others: principles of anthropometry, including the average size of the person who is going to sit in it; principles of economics, including the proportions of the materials to be used; principles of engineering, including the maximum weight of the person who will sit in the chair and the thickness of the material that will support such weight; and principles of ergonomics and health, including the upright position the chair should have and the design of its backrest.

The essence of Love, as the formative cause, showed the author how to justify the origination of the set of notions and the corresponding set of human faculties associated with the formative method of science. It will help us answer the question "How do I envision myself?" by understanding the role of each of the components within the following philosophical argument.

PHILOSOPHICAL ARGUMENT

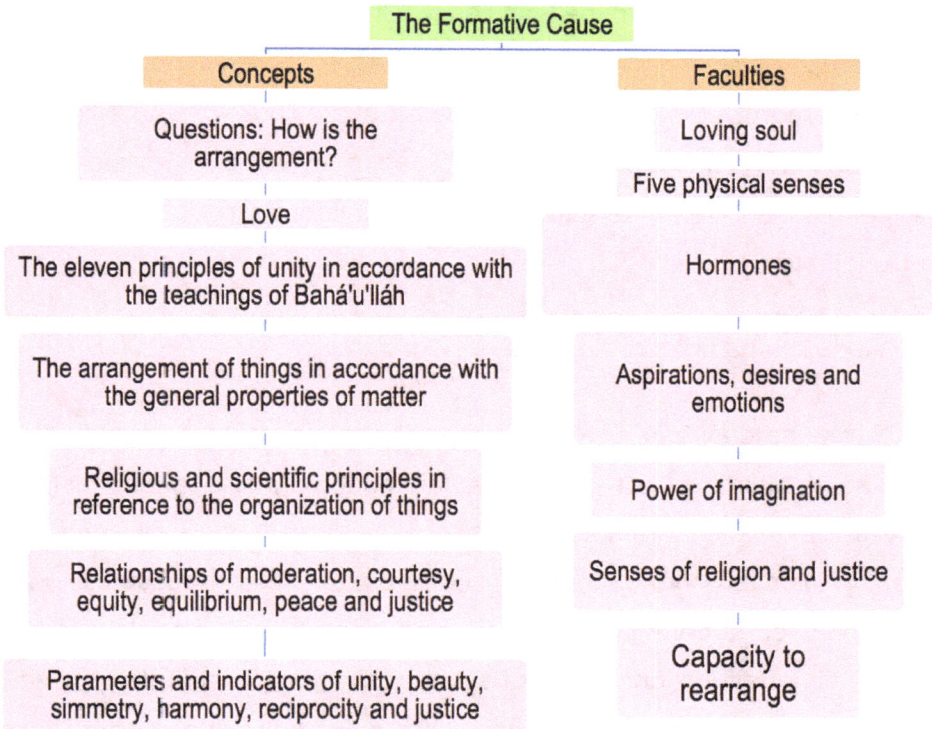

The author has associated the Formative Cause with the question: How? The Formative Cause has been the most difficult question to unravel. Answering it has to do with the form and arrangement of things: How was it arranged? How is it organized now? How are we going to arrange it?

The first approach to solve the Formative Cause and each one of the other causes (Material, Efficient and Final) is to establish the connection between the two quotes by 'Abdu'l-Bahá, that serve as the foundation for this search. In the first quote he mentions the Formal Cause:

> Essential pre-existence is an existence which is not preceded by a cause; essential origination is preceded by a cause. Temporal pre-existence has no beginning; temporal origination has both a

beginning and an end. For the existence of each and every thing depends upon four causes: the Efficient Cause, the Material Cause, *the formal cause*, and the final cause. So this chair has a creator who is a carpenter, a matter which is wood, *a form which is that of a chair*, and a purpose which is to serve as a seat. Therefore, this chair is essentially originated, for it is preceded by, and its existence is conditioned upon, a cause. This is called essential or intrinsic origination.[68]

'Abdu'l-Bahá is reported to have said:

Someone desires an explanation of the terms soul, mind and spirit. The terminology of ancient and modern philosophers differs. According to the great ancient philosophers the words soul, mind and spirit implied the underlying principles of life; the essence was expressed under different names and these three terms designated the various functions of the absolute reality, or the operations of the one single essence; for instance, when they dealt with the *sensations of emotion* they called it the *soul*; when they desired to express that power which discovers the reality of phenomena they gave it the appellation of mind and when they discussed the consciousness which pervades the world of creation they gave it the title of spirit.[69]

After thinking, reading other quotes, the Greek Philosophers, other documents and learning from my own mistakes, I gradually conclude that: the Material Cause and the matter used to make the chair which is wood, most likely was related to the consciousness which pervades the world of

---

[68] 'Abdu'l-Bahá, Some Answered Questions, pp.155 -56. Emphasis added.

[69] 'Abdu'l-Bahá, Divine Philosophy, p. 119. Emphasis added.

creation which they gave the title of spirit; the Efficient Cause could be reasonable connected to that "power which discovers the reality of phenomena which they gave the appellation of mind and the carpenter as the creator of the chair; and, that the sensations of emotion which they called the soul, should be connected to the Formative Cause and the form which is that of a chair. I will make an effort to clarify each one of them.

In the following reflections your will find that form should be linked to the arrangement of things:

Aristotle's hylomorphism concludes that matter receives and supports the form, and form determines matter. In *Human Culture,* Leon Robin says: "… what is interesting to Aristotle on the space is not the vague location of a being within that common and universal place called the world; but the *appropriate place* for that being."[70]

> 'Every string [of the lyre] is *set in the precisely right position* for the due production of the tones within its capacity. . . . The universe is good not when the individual is a stone, but when everyone throws his own voice towards a *total harmony*, singing out a life thin, harsh, imperfect though it be. The *harmony* is made of tones unequal, differing, but *together they form the perfect consonance.*"[71]

But, form does not apply only to material objects. Truthfulness, justice, harmony and unity are moral values and are not perceivable by our senses[72], but, the results of acting in accordance with them can be appreciated with the senses, for example:

---

[70] El Pensamiento Griego y los Orígenes del Espíritu Científico. La Evolución de la Humanidad, p. 264. Emphasis added.

[71] Plotinus, The Six Enneads. Emphasis added.

[72] Abdu'l-Baha, Some Answered Questions, p. 83.

When there is an action leading to the elimination of the extremes of wealth and poverty.

For an action to take place we need a force. A common accepted definition in Physics expresses force, as the result of multiplying mass by acceleration. There are several forces acting simultaneously on any object such as gravity, friction, magnetism. But also: "In the divine world there is love, symbolized in the material world by *magnetism*."[73] "God has willed that love should be a vital *force* in the world, and you all know how I rejoice to speak of love."[74]

But an action or movement to take place needs to have a direction. Social and religious forces have forms, determined by the aspirations, the principles, the motives, the emotions, and the impulses involved, for example: "Let the love of country be superseded by the love of the world."[75] "And if thine eyes be turned towards justice, choose thou for thy neighbor that which thou choosest for thyself."[76]

Love then, is a force but also it is a form, a spiritual quality, a law and an option.

We could assume that the intention of the action, leading to the elimination of the extremes of wealth and poverty, is the direction of the force.

When a jury identifies who is telling the truth in relation to the facts, the truth or is opposite, is the form used by the witnesses to describe the facts of the situation.

---

[73] SOW - Star of the West, Star of the West - 9. Emphasis added.

[74] Abdu'l-Baha, Paris Talks, p. 120. Emphasis added.

[75] SOW - Star of the West, Star of the West - 2

[76] Bahá'u'lláh, Epistle to the Son of the Wolf, p. 29.

When there is harmony and unity within a family because they consult together, consulting within the family is the form used by that existential entity to live in unity and harmony. That direction is the form adopted by the force to achieve the end results.

General Properties of Matter

When human beings organize the arrangement of things, we manage the general properties of matter (form, size, mass, temperature, movement, phases of matter; and position in time and space). In the case of institutions, farms and industries it implies the organization of human and material resources, manufacturing supplies, products and services.

The form, the arrangement and design are much more important than the size to perceive the beauty of justice, harmony and unity. I will call it the "formative method of science", instead of the "quantitative method" in order to keep its true perspective.

Taking agriculture as an example:

> **Form and arrangement**: Which variables fit best the shape of the plots and, the arrangement of the furrows? How about the water drainage and the sun's trajectory?
>
> **Size**: Organization of human resources required in proportion to the size of the plots.
>
> **Mass**: Organization of amount of seed and fertilizer required in proportion to the size of the plots.
>
> **Temperature**: Organization of chores in accordance with the weather forecast.

Position in space: Distribution of the population and the resources. Agricultural zones. The elevation or altitude in meters above sea level. The location of infrastructure, jobs and residences, etc.

Position in time: National, institutional and communal calendars and timetables. Dates of appraisal, project due dates, or market days. Time for planting, or for harvesting. Periods of high or low demand. Periods of scarcity and abundance.

Movement: Utilization of systems of transportation. Attending places of education and others. Administrative activities such as planning monitoring and evaluating. Household chores. Farm and industrial labor. Entertainment and resting. We also organize movements when we vary the speed—i.e., accelerating or decelerating.

Phases of matter: Activities corresponding to climatic seasons. To plan in accordance with the stages of an illness, or the developmental stage of a child.

Religious and Scientific Principles

To understand how a group of people organizes itself to accomplish a task, or what motives are taken into account when an organization distributes its resources, it is useful to look at the relationships among the following: 'the principles of world unity' taught by the Bahá'í Faith, and the organization of humankind; the principles of consultation and team work and the way they organize themselves at the meeting place; the principles of management and the way an institution organizes its resources in order to exercise the monitoring and evaluation of its performance; the principles of economics, education, public health and governance and the organization of society; the principles of environmental preservation and the distribution of natural resources; the principles of engineering and the design of infrastructure, machinery and equipment; the principles of decency and clothing design; the principles of fasting and the environmental temperature during that period[77] or where we work, and so forth. Compare the way the room is arranged when we see the principles of debate being applied and when the principles of consultation are applied, and the outcome in both.

When searching for a solution applying the organizing principles of Science and Religion, we should try to establish the following relationships:

> Religious and scientific principles related to form, organization, design and trajectory

> Religious and scientific principles related to size: length, area, volume, proportion, magnitude

> Religious and scientific principles related to mass

> Religious and scientific principles related to temperature

> Religious and scientific principles related to position in time and in space

> Religious and scientific principles related to phases of matter

> Religious and scientific principles related to movement

> The Formative Cause can be directly associated with the concepts

of structure and model.

---

[77] JE Esslemont, Bahá'u'lláh and the New Era, p. 189.

Love

The fact that 'the principles of world unity' could be related to the organization of the planet, allowed me to understand that the generating force of the Formative Cause is love.

Also, 'Abdu'l-Bahá is reported to have said:

> According to the great ancient philosophers the words soul, mind and spirit implied the underlying principles of life; the essence was expressed under different names and these three terms designated the various functions of the absolute reality, or the operations of the one single essence; for instance, when they dealt with the *sensations of emotion* they called it the soul ....[78]

In truth, the cause of unity is moved by the power of love. There is a relationship between love and affection that is conducive to self-sacrifice. Is it not love that moves us to increase the scope of a health project, or to reach out to vulnerable children? Is it not love that moves us to exert ourselves to polish a part in order to reach the form and excellence desired?

The following quotes express the power of love:

> Know thou of a certainty that Love is the secret of God's holy Dispensation, the manifestation of the All-Merciful, the fountain of spiritual outpourings. Love is heaven's kindly light, the Holy Spirit's eternal breath that vivifieth the human soul. Love is the cause of God's revelation unto man, *the vital bond inherent, in accordance with the divine creation, in the realities of things.* Love is the one means that ensureth true felicity both in this world and the next. Love is the

---

[78] 'Abdu'l-Bahá, Divine Philosophy, p. 119. Emphasis added.

light that guideth in darkness, the living link that uniteth God with man, that assureth the progress of every illumined soul. Love is the most great law that ruleth this mighty and heavenly cycle, the unique power that bindeth together the divers elements of this material world, the supreme magnetic force that directeth the movements of the spheres in the celestial realms. Love revealeth with unfailing and limitless power the mysteries latent in the universe. Love is the spirit of life unto the adorned body of mankind, the establisher of true civilization in this mortal world, and the shedder of imperishable glory upon every high-aiming race and nation.[79]

All things are beneficial if joined with the love of God; and without His love all things are harmful, and act as a veil between man and the Lord of the Kingdom. When His love is there, every bitterness turneth sweet, and every bounty rendereth a wholesome pleasure. For example, a melody, sweet to the ear, bringeth the very spirit of life to a heart in love with God, yet staineth with lust a soul engrossed in sensual desires. And every branch of learning, conjoined with the love of God, is approved and worthy of praise; but bereft of His love, learning is barren -- indeed, it bringeth on madness. Every kind of knowledge, every science, is as a tree: if the fruit of it be the love of God, then is it a blessed tree, but if not, that tree is but dried-up wood, and shall only feed the fire.[80]

---

[79] 'Abdu'l-Bahá, Selections, p. 27. Emphasis added.

[80] ibid. p. 181.

The Soul

The Master, 'Abdu'l-Bahá, is reported to have said that when the great philosophers of the past "dealt with the sensations of emotion they called it the *soul*"[81], which is why I also have associated it with love. "If we are caused joy or pain by a friend, if a love prove true or false, it is the *soul* that is affected. If our dear ones are far from us -- it is the *soul* that grieves, and the grief or trouble of the *soul* may react on the body. …The *soul* does not evolve from degree to degree as a law -- it only evolves nearer to God, by the Mercy and Bounty of God."[82]

Elsewhere we find:

> The human spirit, which distinguishes man from the animal, is the rational soul, and these two terms—the human spirit and the rational soul—designate one and the same thing. This spirit, which in the terminology of the philosophers is called the rational soul, encompasses all things and as far as human capacity permits, discovers their realities and becomes aware of the properties and effects, the characteristics and conditions of earthly things. But the human spirit, unless it be assisted by the spirit of faith, cannot become acquainted with the divine mysteries and the heavenly realities. It is like a mirror which, although clear, bright, and polished, is still in need of light. Not until a sunbeam falls upon it can it discover the divine mysteries.[83]

---

[81] Divine Philosophy, p. 120.

[82] 'Abdu'l-Bahá, Paris Talks, p. 729. Emphasis added.

[83] 'Abdu'l-Bahá, Some Answered Questions, p. 55.

Know, verily, that the soul is a sign of God, a heavenly gem whose reality the most learned of men hath failed to grasp, and whose mystery no mind, however acute, can ever hope to unravel. It is the first among all created things to declare the excellence of its Creator, the first to recognize His glory, to cleave to His truth, and to bow down in adoration before Him. If it be faithful to God, it will reflect His light, and will, eventually, return unto Him. If it fail, however, in its allegiance to its Creator, it will become a victim to self and passion, and will, in the end, sink in their depths.[84]

It has been reported that the Master, 'Abdu'l-Bahá, said:

If the soul identifies itself with the material world it remains dark, for in the natural world there is corruption, aggression, struggles for existence, greed, darkness, transgression and vice. If the soul remains in this station and moves along these paths it will be the recipient of this darkness; but if it becomes the recipient of the graces of the world of mind, its darkness will be transformed into light, its tyranny into justice, its ignorance into wisdom, its aggression into loving kindness; until it reach the apex. Then there will not remain any struggle for existence. Man will become free from egotism; he will be released from the material world; he will become the personification of justice and virtue, for a sanctified soul illumines humanity and is an honor to mankind, conferring life upon the children of men and suffering all nations to attain to the station of perfect unity. Therefore, we can apply the name 'holy soul' to such a one.[85]

---

[84] Bahá'u'lláh, Gleanings, p. 158.

[85] 'Abdu'l-Bahá, Divine Philosophy, p. 121.

Relationships

In general, the aforementioned religious and scientific principles can be used to frame relationships of moderation, courtesy, equity, unity, correspondence, equilibrium, harmony, love, peace and justice. Within them is implicit an organization of things.

The laws that are related to the Formative Cause are the law of love, the laws that regulate relationships, the law of gravity, and the law of attraction of masses.

Parameters and Indicators

One of the roles of governments, managers of enterprises and mature individuals is to be able to monitor and evaluate the efforts made in pursuit of their vision. We should establish parameters and indicators to monitor and evaluate the advancement in the achievement of beauty, justice, unity, decency and harmony in the context of the adopted organization or arrangement.

Applying the Single Science to the Formative Cause

The table below shows the methodology followed in the construction of the "New Approach to Science" by inter-weaving the causes horizontally and vertically. In order for the reader to understand the placement of the following acceptations in one of the Rows (1 to 6), it is important to understand how the Formative Cause is interwoven with the other causes.

	THE FORMATIVE CAUSE AND THE FORMATIVE METHOD OF SCIENCE:	
1	Question: How is the arrangement?	MATURITY CAUSE
2	**Essence**: love and soul as its power. **The general properties of matter**: form, size, mass, temperature, movement and position in time and space **Hormones, feelings and emotions**	MATERIAL CAUSE
3	**Laws** related to unity, harmony, justice and the law of gravity. **Goal**: Aspirations and desires to reach the expected vision or design	FINAL CAUSE
4	**Powers**: imagination and senses of justice and religion **The practical application of the principles related to the general properties of matter**: form, size, mass, temperature, movement and position in time and space in relation to the desire organization, arrangement and design	EFFICIENT CAUSE
5	**Religious and Scientific principles related to the general properties of matter**	FORMATIVE CAUSE
6	**Parameters and indicators** of unity, beauty, symmetry, harmony, reciprocity and justice in reaching the desire vision or design	MATURITY CAUSE

Having the basis of a conceptual framework for the Formative Cause and learned in the single science proposed above in "Classes and Categories as foundation of Philosophy", that we are supposed to group each notion of interest for the solution of a problem in one of the causes; the following sample of a general accepted meaning of a word could be placed in the Formative Cause:

Meristem: "A formative plant tissue usually made up of small cells capable of dividing indefinitely and giving rise to similar cells or to cells that differentiate to produce the definitive tissues and organs."[86] (In row 4 the intersection of The Formative Cause and The Efficient Cause because is describing the meristem's function in the formation of a tissue, but, it also could be placed in row 1 the intersection of The Formative Cause and The Cause of Maturity because it is arranged to involuntarily generate similar cells or differentiated ones to form the definitive tissues and organs, but, could also be placed in row 5 the intersection of The Formative Cause and The Formative Cause because *similar* and *differentiated* are referring to the form of the cells and small is referring to size, which are general properties of matter)

Abscisic: "Abscisic acid A plant hormone inhibiting growth; helps plants withstand adverse conditions."[87] (In row 5 The Formative Cause and The Formative Cause because it is referring to size, but, could also be placed in row 1 the intersection of The

---

[86] "meristem." <https://quizlet.com>.

[87] "abscisic." <https://quizlet.com>.

Formative Cause and The Maturity Cause because it helps plants withstand adverse conditions)

Auxin: "Any of various hormones or similar substances that promote and regulate the growth and development of plants. Auxins are produced in the meristem of shoot tips and move down the plant, causing various effects."[88] (In row 5 the intersection of The Formative Cause and The Formative Cause because it is referring to size, but, also could be placed in row 1 the intersection of The Formative Cause and The Cause of Maturity because it involuntary regulate its growth, or even better expressed is the Will of God, but, could also be placed in row 4 the intersection of The Formative Cause and The Efficient Cause because Auxins are *produced* in the meristem of shoot tips and *move* down the plant, causing various *effects*)

Synchrony: "A state in which things happen, move, or exist at the same time."[89] (In row 2 the intersection of The Formative Cause and The Material Cause" because if *things exist* at the same time we should be able to perceive their existence properties, but, also in row 4 the intersection of The Formative Cause and The Efficient Cause because it is an state in which things move or exist at the same time, but, it also could be placed in row 5 "the intersection of The Formative Cause and The Formative Cause" because time is a general property of matter; and, could also be placed in row 6 the intersection of

---

[88] "auxin." <https://quizlet.com>.

[89] "synchrony." Merriam-Webster Dictionary.

The Formative Cause and "The Cause of Maturity" because synchrony can be used as a parameter)

Hormone: "Circulating molecules that serve as signals for particular body processes to occur by interacting with target cells. However, there are some specific hormones that greatly affect human emotions. These hormones include Estrogen, Progesterone, Testosterone, Norepinephrine and Epinephrine, Serotonin, GABA, Dopamine, Acetylcholine, and Oxytocin."[90] (In row 2 the intersection of The Formative Cause and The Material Cause because hormones are molecules, but, it also could be placed in row 5 the intersection of The Formative Cause and The Formative Cause because some specific *hormones* greatly affect human *emotions*)

Synthesis: "Something that is made by combining different things (such as ideas, styles, etc.). The production of a substance by combining simpler substances through a chemical process."[91] (In row 2 the intersection of The Formative Cause and The Material Cause because it is defined as the combination of different things and because of it, can be assimilated to unity in diversity which is one of the principles of unity, but, also could be placed in row 4 the intersection of The Formative Cause and The Efficient Cause because it is the *production* of a substance by combining simpler substances through a chemical *process* or made by combining different things)

---

[90] "hormone." Eugene M. Mccarthy, Online Biology Dictionary.

[91] "synthesis." Merriam-Webster Dictionary.

Of course, as in the case of the chair, each one of those words can be perceived from different perspectives. What is the potential of being able to classify those acceptations in this cause?

The same grouping criteria applies to the below mentioned faculties associated with the Formative Cause.

OTHER HUMAN FACULTIES LINKED TO THE FORMATIVE CAUSE

The human faculties described further down can help us answer the question: How do I react to injustice, discrimination, oppression, disunity and similar situations?

Besides having a soul and being able to exercise the power of love, human beings have been endowed with:

Hormones and Emotions

Our bodies respond differentially to stimulus with feelings and emotions and their relationship with hormones and the endocrine system.

Power of Imagination

After evaluating the past and the current arrangement or organization we have to imagine diverse scenarios in order to reach the desire outcome.

Imagination is a very important instrument of creativity, but it requires the proof of science. Focusing on the power of imagination, let us consider the following quotes: "Man has likewise a number of spiritual powers: *the power of imagination, which forms a mental image of things*; thought, which reflects upon the realities of things; comprehension, which understands these realities; and memory, which retains whatever man has imagined, thought, and understood."[92]

> If religious beliefs and opinions are found contrary to the standards of science they are mere superstitions and *imaginations*; for the antithesis of knowledge is ignorance, and the child of ignorance is superstition. Unquestionably there must be agreement between true religion and science. If a question be found contrary to reason, faith and belief in it are impossible and there is no outcome but wavering and vacillation.[93]

Sense of Religion

By acknowledging the existence of the sense of religion, our faith in reaching the promises of a golden age for humanity is augmented. "Taken in general, women today have a stronger *sense of religion* than men. The woman's intuition is more correct; she is more receptive and her intelligence is quicker. The day is coming when woman will claim her superiority to man."[94]

How could we expect to attain "the principles of world unity" without enhancing our sense of religion and our sense of justice?

Sense of Justice

If the following quotation is true:

---

[92] Some Answered Questions, p. 56. Emphasis Added.

[93] 'Abdu'l-Bahá, The Promulgation of Universal Peace, p. 181. Emphasis Added.

[94] 'Abdu'l-Bahá,'Abdu'l-Bahá in London, p. 104. Emphasis Added.

O SON OF SPIRIT! The best beloved of all things in My sight is *Justice*; turn not away therefrom if thou desirest Me, and neglect it not that I may confide in thee. By its aid thou shalt see with thine own eyes and not through the eyes of others, and shalt know of thine own knowledge and not through the knowledge of thy neighbor. Ponder this in thy heart; how it behooveth thee to be. Verily justice is My gift to thee and the sign of My loving-kindness. Set it then before thine eyes.[95]

Why should man, who is endowed with the *sense of justice* and sensibilities of conscience, be willing that one of the members of the human family should be rated and considered as subordinate? Such differentiation is neither intelligent nor conscientious; therefore, the principle of religion has been revealed by Bahá'u'lláh that woman must be given the privilege of equal education with man and full right to his prerogatives.[96]

Of course, comprehending that religion is progressive we can examine the validity of God's teachings in different epochs, for example:

On the societal level, the principle of collective security enunciated by Bahá'u'lláh (see Gleanings from the Writings of Bahá'u'lláh, CXVII) and elaborated by Shoghi Effendi (see the Guardian's letters in The World Order of Bahá'u'lláh) does not presuppose the abolition of the use of force, but prescribes 'a system in which Force is made the servant of Justice', and which provides for the existence of an international peace-keeping force that 'will safeguard the organic

---

[95] Bahá'u'lláh, The Hidden Words of Bahá'u'lláh. Emphasis Added.

[96] 'Abdu'l-Bahá, The Promulgation of Universal Peace, p. 106. Emphasis added.

unity of the whole commonwealth. In the Tablet of Bishárát, Bahá'u'lláh expresses the hope that 'weapons of war throughout the world may be converted into instruments of reconstruction and that strife and conflict may be removed from the midst of men'.

In another Tablet Bahá'u'lláh stresses the importance of fellowship with the followers of all religions; He also states that 'the law of holy war hath been blotted out from the Book'.[97]

*

With this set of human faculties, that is common to all human beings, we get an idea of the beauty of the setting or the structure.

In conclusion, the cycle of scientific research has also a formative phase. The notions and the set of human faculties associated in the Formative Cause are appropriate to be able to answer questions such as: How are we organized? How do we envision our lives?

**

In his search for a single science Aristotle wrote:

The minute accuracy of *mathematics* is not to be demanded in all cases, but only in the case of things which have no matter. Hence method is not that of natural science; for presumably the whole of nature has matter. Hence we must inquire first what nature is: for thus we shall also see what *natural science* treats of (and whether it *belongs to one science or to more to investigate the causes and the principles* of things).[98]

---

[97] Bahá'u'lláh. The Kitáb-i-Aqdas, p. 241, Note 173.

[98] The Metaphysics. Emphasis added.

Further down he wrote:

> ... some do not think there is anything substantial besides sensible things, but others think there *are eternal substances* which are more in number and more real; e.g. Plato posited two kinds of substance-the *Forms and objects of mathematics*-as well as a third kind, viz. *the substance of sensible bodies*. And Speusippus made still more kinds of substance, beginning with the One, and assuming principles for each kind of substance, one for *numbers*, another for *spatial magnitudes*, and then another for the *soul*, and by going on in this way he multiplies the kinds of substance. And some say *Forms and numbers* have the same nature, and the other things come after them-*lines and planes*-until we come to the substance of the material universe and to *sensible bodies*.[99]

Let us read again what the wise Diotima taught Socrates about the meaning of love in connection to the words where emphasis has been added:

> For he who would proceed aright in this matter should begin in youth to visit *beautiful forms*; and first, if he be guided by his instructor aright, *to love one such form* only-out of that he should create *fair thoughts*; and soon he will of himself perceive that *the beauty* of one *form* is akin to the *beauty* of another; and then *if beauty of form* in general is his pursuit, how foolish would he be not to recognize that the *beauty in every form* is and the same! And when he perceives this he will abate his violent *love of the one*, which he will despise and deem a small thing, and will become *a lover of all beautiful forms* ... until he is compelled to contemplate and see the *beauty* of institutions

---

[99] Ibid.

and laws, and to understand that the *beauty* of them all is of one family, and that personal *beauty* is a trifle; and after laws and institutions he will go on to the sciences, that he may see their *beauty*, being not like a servant in *love with the beauty* of one youth or man or institution, himself a slave mean and narrow-minded, but drawing towards and contemplating the vast sea of beauty, he will create many *fair* and noble thoughts and notions in *boundless love of* wisdom; until on that shore he grows and waxes strong, and at last the *vision* is revealed to him of a *single science*, which is the science of *beauty everywhere*.[100]

---

[100] Plato, Symposium. Emphasis added.

VI THE FINAL CAUSE AND THE EXPLICATIVE METHOD OF SCIENCE

In order to introduce the "Final Cause" let us advance with the example of the chair. The *"Why?"* and *"For what?"* of the chair are determined by its purpose, in other words, the purpose of its construction. When they decided to mention such a thing, that was to be used as a seat, they agreed to call it a chair. A living room chair is supposed to be indoors and is not supposed to be exposed to the weather's inclemency. It also must provide safety to whoever rests in it.

The essence of the Laws of God, as the final cause, put the author in the right track to articulate the set of notions and the corresponding set of human faculties associated with the Explicative method of science. It will help us to answer the questions: "Why?" and "For what do I exist?"

PHILOSOPHICAL ARGUMENT

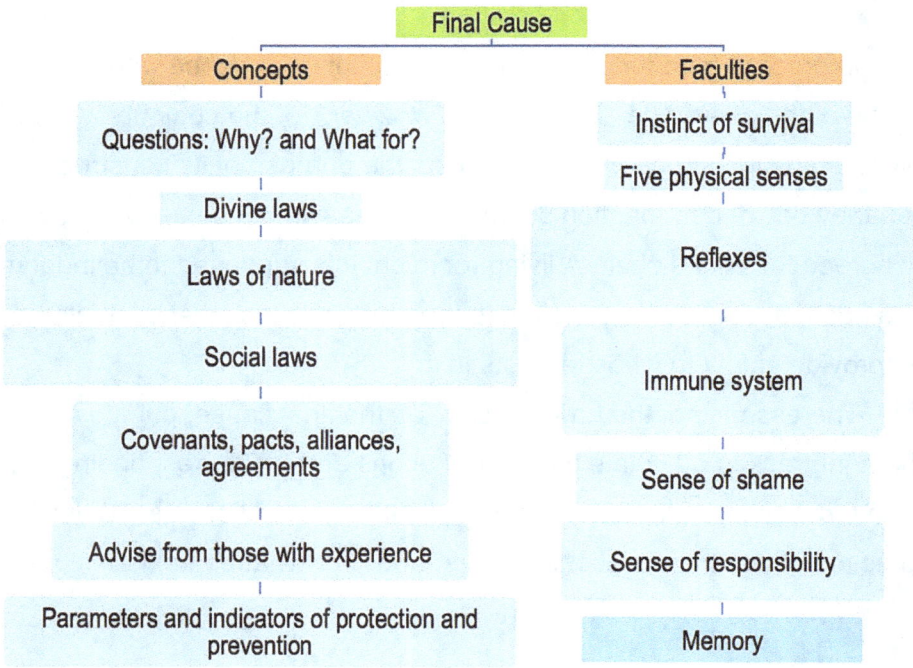

The final cause has been linked by philosophers with concepts such as benefit, aim, purpose, and ideal. The final cause is defined as the reason why something was created or made. The final cause is associated with the questions: "Why?" and "For what?". It is used in strategic planning exercises to determine the institutional mission.

When we reflect on the questions "Why?" and "For what?" it is difficult to relate them to concepts other than law. Let us carefully study the following extract of the dialogue 'Crito' / Or the Duty of a Citizen to understand the relationship between the final cause and laws.

Socrates having been condemned to death, his friend Crito visits him and says:

Crito. So let it be, then. But answer me this, Socrates, are you not anxious for me and other friends, lest, if you should escape from hence, informers should give us trouble, as having secretly carried you off, and so we should be compelled either to lose all our property, or a very large sum, or to suffer something else besides this? For, if you fear anything of the kind, dismiss your fears; for we are justified in running the risk to save you—and, if need be, even a greater risk than this. But be persuaded by me, and do not refuse.

Socrates. I am anxious about this, Crito, and about many other things.[101]

Further down, we find:

Socrates. My dear Crito, your zeal would be very commendable were it united with right principle; otherwise, by how much the more earnest it is, by so much is it the more sad. We must consider, therefore, whether this plan should be adopted or not. For I not now only, but always, am a person who will obey nothing within me but reason, according as it appears to me on mature deliberation to be best. And the reasons which I formerly professed I cannot now reject, because this misfortune has befallen me; but they appear to me in much the same light, and I respect and honor them as before; so that if we are unable to adduce any better at the present time, be assured that I shall not give in to you, even though the power of the multitude should endeavor to terrify us like children, by threatening more than it does now, bonds and death, and confiscation of property. How, therefore, may we consider the matter most conveniently? First of all, if we recur to the argument which you used about opinions, whether on former

---

[101] Plato, Crito: or, the Duty of a Citizen, p. 4.

occasions it was rightly resolved or not, that we ought to pay attention to some opinions, and to others not; or whether, before it was necessary that I should die, it was rightly resolved; but now it has become clear that it was said idly for argument's sake, though in reality it was merely jest and trifling. I desire then, Crito, to consider, in common with you, whether it will appear to me in a different light, now that I am in this condition, or the same, and whether we shall give it up or yield to it. It was said, I think, on former occasions, by those who were thought to speak seriously, as I just now observed, that of the opinions which men entertain some should be very highly esteemed and others not. By the gods! Crito, does not this appear to you to be well said? For you, in all human probability, are out of all danger of dying to-morrow, and the present calamity will not lead your judgment astray. Consider, then; does it not appear to you to have been rightly settled that we ought not to respect all the opinions of men, but some we should, and others not? Nor yet the opinions of all men, but of some we should, and of others not? What say you? Is not this rightly resolved?

Crito. It is.

Socrates. Therefore we should respect the good, but not the bad?

Crito. Yes.

Socrates. And are not the good those of the wise, and the bad those of the foolish?

Crito. How can it be otherwise?[102]

Further down, we find:

Socrates. Observe, then, what follows. By departing hence without the leave of the city, are we not doing evil to some, and that to those to whom we ought least of all to do it, or not? And do we abide by what we agreed on as being just, or do we not?

Crito. I am unable to answer your question, Socrates; for I do not understand it.

Socrates. Then, consider it thus. If, while we were preparing to run away, or by whatever name we should call it, the laws and commonwealth should come, and, presenting themselves before us, should say, 'Tell me, Socrates, what do you purpose doing? Do you design any thing else by this proceeding in which you are engaged than to destroy us, the laws, and the whole city, so far as you are able? Or do you think it possible for that city any longer to subsist, and not be subverted, in which judgments that are passed have no force, but are set aside and destroyed by private persons?'—what should we say, Crito, to these and similar remonstrances? For any one, especially an orator, would have much to say on the violation of the law, which enjoins that judgments passed shall be enforced. Shall we say to them that the city has done us an injustice, and not passed a right sentence? Shall we say this, or what else?

---

[102] ibid. p. 6.

Crito. This, by Jupiter! Socrates.[103]

Socrates. What, then, if the laws should say, 'Socrates, was it not agreed between us that you should abide by the judgments which the city should pronounce?' And if we should wonder at their speaking thus, perhaps they would say, 'Wonder not, Socrates, at what we say, but answer, since you are accustomed to make use of questions and answers. For, come, what charge have you against us and the city, that you attempt to destroy us? Did we not first give you being? and did not your father, through us, take your mother to wife and beget you? Say, then, do you find fault with those laws among us that relate to marriage as being bad?' I should say, 'I do not find fault with them.' 'Do you with those that relate to your nurture when born, and the education with which you were instructed? Or did not the laws, ordained on this point, enjoin rightly, in requiring your father to instruct you in music and gymnastic exercises?' I should say, rightly. Well, then, since you were born, nurtured, and educated through our means, can you say, first of all, that you are not both our offspring and our slave, as well you as your ancestors? And if this be so, do you think that there are equal rights between us? and whatever we attempt to do to you, do you think you may justly do to us in turn? Or had you not equal rights with your father, or master, if you happened to have one, so as to return what you suffered, neither to retort when found fault with, nor, when stricken, to strike again, nor many other things of the kind; but that with your country and the laws you may do so; so that if we attempt to destroy you, thinking it to be just, you also should endeavor, so far as you are able, in return, to destroy us, the

---

[103] ibid. p. 11.

laws, and your country; and in doing this will you say that you act justly—you who, in reality, make virtue your chief object? ... .[104]

The argument continues until the end of the Dialogue, highlighting the transcendence of the laws and the obedience they require from the citizens. Socrates also emphasizes the importance of obedience to the Republic, even after it has mistakenly condemned him to death. Socrates perceived his material existence as something ephemeral, and was willing to sacrifice it in order to illuminate the world.

Having appreciated the importance of laws, the question that immediately arises is: are they inherent to all things? Western civilization has taught us that social laws and religious laws are external or extrinsic to human beings. But this is not the case. Consider the following quotes, which refer to Divine Laws:

> The laws and ordinances that constitute the major theme of this Book, Bahá'u'lláh, moreover, has specifically characterized as '*the breath of life unto all created things*', as 'the mightiest stronghold', as the 'fruits' of His 'Tree', as 'the highest means for the maintenance of order in the world and the security of its peoples', as 'the lamps of His wisdom and loving-providence', as 'the sweet-smelling savour of His garment', and the 'keys' of His 'mercy' to His creatures. 'This Book', He Himself testifies, 'is a heaven which We have adorned with the stars of Our commandments and prohibitions'. ... 'Say, O men! Take hold of it with the hand of resignation ... By My life! It hath been sent down in a manner that amazeth the minds of men. Verily, it is My

---

[104] ibid. p. 12.

weightiest testimony unto all people, and the proof of the All-Merciful unto all who are in heaven and all who are on earth.[105]

And:

Blessed the palate that savoureth its sweetness, and the perceiving eye that recognizeth that which is treasured therein, and the understanding heart that comprehendeth its allusions and mysteries. By God! Such is the majesty of what hath been revealed therein, and so tremendous the revelation of its veiled allusions that the loins of utterance shake when attempting their description.' And finally: 'In such a manner hath the Kitáb-i-Aqdas been revealed that it attracteth and embraceth all the divinely appointed Dispensations. Blessed those who peruse it! Blessed those who apprehend it! Blessed those who meditate upon it! Blessed those who ponder its meaning! *So vast is its range that it hath encompassed all men ere their recognition of it.* Erelong will its sovereign power, *its pervasive influence and the greatness of its might be manifested on earth.*[106]

Later: "Say: This is the very soul of all Scriptures which hath been breathed into the Pen of the Most High, causing all created beings to be dumbfounded, save only those who have been enraptured by the gentle breezes of *My loving-kindness and the sweet savours of My bounties which have pervaded the whole of creation.*"[107]

---

[105] Shoghi Effendi qtd. in the Introduction of The Most Holly Book by Bahá'u'lláh. The Kitáb-i-Aqdas, pp. 15-16. Emphasis Added.

[106] ibid. p. 16. Emphasis Added.

[107] Bahá'u'lláh, The Kitáb-i-Aqdas, p. 68. Emphasis Added.

The above is the reason why the author decided to choose "the Laws of God" as the essential origination of the Final Cause.

Within the philosophy of certain indigenous populations, such as the Arhuacos in Colombia, the violation of a social law is also considered as against your own nature. Indeed, accepting the challenge of attempting to establish harmony between social laws and the laws of God, whether they are manifested in nature or in the Sacred Books revealed by Him, allows us to comprehend why we sometimes feel that laws are imposed on us, as if they were external to ourselves; but once we start grasping their inherent wisdom, our love for them increases.

Order in a country, and in any community, is dependent on the respect for its constitution, social laws, natural laws, and obedience to those in authority. The relationships of what is known as the Golden Rule of all religions, require us to examine the knowledge that we possess about laws and concepts such as the power of God and the institutions He ordained, the power of human institutions, rights and duties, protection, freedom, self-determination, obedience, authority, and responsibility.

The flourishing of the potentialities and powers of all things depends, ultimately, on their submission to the laws and norms by which they are governed. For human beings, this recognition must be reached willingly, through an understanding of the reasons that underpin it and through "love for God as the motive of obedience to His Laws."[108]

It is in this recognition that the causes of the deepest problems must be sought. This is the key for establishing the general purposes of any institution, government or individual.

The information about the final cause can be systematized using natural laws, religious laws and ordinances, social norms,

---

[108] Shoghi Effendi qtd. in Synopsis and Codifications of the Laws and Ordinances of The Kitáb-i-Aqdas in Bahá'u'lláh, The Kitáb-i-Aqdas, p. 163.

recommendations, risks, consequences, relationships of causality, consequences, parameters and indicators of prevention and protection.

As a very important input to enrich the concept of human being in the Final Cause is the understanding of what citizenship means. William Huitt expresses:

> Discussions regarding citizenship have received extensive attention in recent years as humanity moves rapidly into a new era of globalization (e.g., Isin, 2000; King, 2000; Peters, Britton, & Blee, 2008; Roth & Burbules, 2007).
>
> There are at least three issues that should be of concern to educators: identity, loyalty and responsibility, and rights.
>
> In many ways, one's concept of citizenship is an essential element of one's self identity. At the earliest stages of human evolution, the group affiliation that provided a source of one's identity was the family or band, which then evolved into tribe, city state, and empire (McNeill & McNeill, 2003). These changes in affiliation and identity took tens of thousands, then thousands, then hundreds of years. While most people had an identity at only one of these levels at any given time, the concept of world citizen (derived from the Greek word kosmopolitês) had its advocates even in antiquity (Kleingeld, 2006). The geographical explorations of the world in the fifteenth to seventeenth centuries (McNeill, & McNeill, 2003), the rapid changes in technology in the nineteenth and twentieth centuries (Huitt, 2007), combined with an increased diaspora that is expected to continue (Castles & Miller, 2003; OECD, 2008), has created a complexity of affiliation and identity seen only in isolated individuals in the past (Banks, 2007; Grimshaw & Sears, 2008; Marshall, 2009). Unfortunately, recent attempts to make sense of these changes result

in contradictory views. One such example is found in the statements that the world is flat (Friedman, 2007) and the world is curved (Smick, 2008). Townsend (2009) argued that this particular contradiction results from a change in focus--either on the whole (flat) or the differentiation of the parts (curved)--and that in the postmodern world, people (especially leaders) need to think and act at both levels and all of those in between. Abrams and Primack (2011) make the case that one's identity should have a relationship to the cosmos as each human being, at least his or her material form, is a direct result of the evolution of the cosmos. This is the perspective taken by Brown (2007) and Christian (2005) in their development of the concept of big history. A major advantage for those advocating a local-global mindset (Bell-Rose & Desan, 2006; Townsend, 2009) is that people are able to understand one level above where they are able to act (Perry, 1999; Reimer, Paolitto, & Hersch, 1983). Developing an identity as a global citizen will be easier for those who identify themselves as citizens of the cosmos.

Another issue involves the contradictions of loyalty (Hansen, 2010) and responsibility (Karlberg, 2008). On the one hand, individuals owe certain loyalties and have responsibilities to their communities, as they are the primary contexts with which individuals have daily contact (Shinn & Toohey, 2003). In fact, it is neighborhoods that are the first level of community and serve as a developmental context, especially for children in poverty (Vaden-Kierman, D'Elio, O'Brien, Tarullo, Zill, and Hubbell-McKey, 2010). However, at least in the USA, individuals are becoming more isolated and detached from their neighborhoods and immediate geographical regions (Putnam, 2000). Turkle (2012) suggested this is a direct result of being tethered digitally throughout every waking moment of one's life. With the spread of wireless technology and the use of smartphones and other mobile devices throughout the world (Corasaniti, 2010; International Telecommunications Union, 2011), this phenomenon is likely to increase. While it is advocated that people spend more time "untethered" (Bourg Carter, 2012), perhaps having a sense of loyalty to the cosmos as a cosmic citizen will encourage people to develop their full potentials as a means of being loyal to the cosmos into which they were born and using their competencies in the development of their local communities.

The reality is, however, that in the modern world, the nation-state is the focus of one's loyalties and responsibilities (Koczanowicz, 2010). At the same time, the populations of the nation-state across the world are morphing rapidly as there is an unprecedented number of foreign-born individuals within a specific nation state (OECD, 2009). This is putting tremendous pressure on nation states to both prepare their citizens to interact with and perhaps live and work in other countries while at the same time integrating non-native people into the society (Koczanowicz, 2010). To make matters even more complex, the various relationships are nested (i.e., individual within local community within province, state, or region within nation within international region within world), and there are reciprocal (i.e., back-and-forth) relationships at all levels (Huitt, 2012a). The fluidity of people's movements in, out, and through these various relationships is unprecedented in human history, at least on a global basis, making it very difficult to define one's loyalties and responsibilities.

Another contradiction is the discussion of whose rights should be central to the concept of citizenship: that of the individual (Hall, Coffey, & Williamson, 1999) or that of the community (Stevenson, 2010). While there are excellent rationales provided for both of these as a focus for identifying rights of citizens, there is also an advocacy that the most important consideration is to provide a dynamic balance between the perspectives of individual autonomy and collective benefit (McIntyre-Mills, 2009). This theme is adopted in the United Nations (1948) Universal Declaration of Human Rights. Pykett (2010) suggested that discussing the tensions between individual freedoms and social order is crucial to developing a sustainable view of citizenship education and guiding social reforms. At a time in history when society is in great flux, a lack of a coherent policy results in jumping back and forth between these two advocacies in a manner that is neither satisfying nor effective. Again, having an identity as a cosmic citizen can impact the development of a concept of human rights from a global perspective.[109]

Bahá'u'lláh in reference to citizenship stated:

It is incumbent upon every man of insight and understanding to strive to translate that which hath been written into reality and action.... That one indeed is a man who, today, dedicateth himself to the service of the entire human race. The Great Being saith: Blessed and happy is he that ariseth to promote the best interests of the peoples and kindreds of the earth. In another passage He hath proclaimed: It is not for him to pride himself who loveth his own country, but rather for him who loveth the whole world. The earth is but one country, and mankind its citizens.[110]

---

[109] Citizenship. Cosmic-Citizenship.

[110] Gleanings from the Writings of Bahá'u'lláh, p. 250.

Relationships

When considering "Why?" and "For what?" the relationships are of: gratitude, loyalty, reciprocity, mutuality and those of fear and protection; and also between the individual and the law and the institutions such as: guilt, repentance, punishment and reward.

Parameters and Indicators of Protection, Equality, Prevention and Security

Parameters and indicators that can be used to monitor and evaluate the situation such as: those that guarantee that the law is for all, those to measure the handling of aggressive behavior, those used to guarantee food security and food safety for all and indicators such as: Proportion of inhabitants vaccinated.

Applying the Single Science to the Final Cause

The table below shows the methodology followed in the construction of the "New Approach to Science" by inter-weaving the causes horizontally and vertically. In order for the reader to understand the placement of the following acceptations in one of the Rows (1 to 6), it is important to understand how the Final Cause is interwoven with the other causes.

	THE FINAL CAUSE AND THE EXPLICATIVE METHOD OF SCIENCE	
1	**Questions:** Why? For what?	**CAUSE OF MATURITY**
2	**Essence:** the power of law itself, i.e., that which is a condition to a certain aspect of life or nature, and explains its purpose, its mission	**MATERIAL CAUSE**
3	Social order, laws and norms; Natural laws; Religious laws and ordinances. Pacts, agreements and covenants. Individual and collective commitment to ideals **Goal:** the end, the mission, the purpose	**FINAL CAUSE**
4	**Powers:** Memory and senses of responsibility, fear and shame **Advice** from those with experience. The **consequences** of obeying and disobeying.	**EFFICIENT CAUSE**
5	**Situations** of risk and danger. **Relationships** of gratitude, loyalty, reciprocity, mutuality, of fear and protection; and also guilt, repentance, punishment and reward.	**FORMATIVE CAUSE**
6	**Parameters and indicators of protection, equality, prevention and security** in the fulfillment of the mission	**CAUSE OF MATURITY**

Having the basis of a conceptual framework for the Final Cause and learned in the single science proposed above in "Classes and Categories as foundation of Philosophy", that we are supposed to classify each notion of interest for the research in one of the causes; the following sample of a usual accepted meaning of a word could be placed in the Final Cause:

Mutualism: "A form of symbiosis in which both participants benefit. For example, a clown fish lives inside a sea anemone and is protected by it. In return, it brings scraps to the anemone, and lures larger fish into the anemone's tentacles."[111] (In row 3 the intersection of The Final Cause and The Final Cause because it can be assimilated to an involuntary pact, but, could also be placed in row 5 the intersection of The Final Cause and The Formative Cause because Mutualism is a *form* of symbiosis)

Infection: "Infection is the process or the state wherein an infectious agent (such as pathogenic microorganisms, viruses, prions, viroids, nematodes and helminths) invades and multiplies in the body tissues of the host. It could result in the manifestation of symptoms and disease when the immune response of the host is activated. It may also be palpable when the infection results in the competition for nutrients and metabolism."[112] (In row 4 the intersection of The Final Cause and The Efficient Cause because "Infection is the *process* or the *state* wherein an infectious agent" may "*result* in the *manifestation* of symptoms and disease", but, also in row 5 the intersection of The Final Cause and The Formative Cause because it is referring to size when an infectious agent invades and *multiplies* in the body tissues of the host becoming a situation of danger, but, could also be placed in row 3 the intersection of

---

[111] "mutualism." Eugene M. Mccarthy, Online Biology Dictionary.

[112] "infection." Biology Online.

The Final Cause and The Final Cause because the possible infectious *consequences* affecting the purpose of the body tissues of the host)

Toxin: "A poison, produced by an animal or plant, that elicits the production of an antibody (antitoxin) when introduced into bodily tissue, typically by injection."[113] (In row 4 the intersection of The Final Cause and The Efficient Cause because it is a poison, *produced* by an animal or plant, but, also could be placed in row 6 the intersection of The Final Cause and The Cause of Maturity because it is an involuntary differential response to elicit the production of an *antitoxin* preventing further damage , but, could also be placed in row 3 the intersection of The Final Cause and The Final Cause because its *poisonous* effect becoming a situation of risk)

Langerhans cell: "A type of dendritic cell found in epidermis. As part of the epidermal immune system, these cells act as antigen-presenting cells."[114] (In row 2 the intersection of The Final Cause and The Material Cause because an antigen is "any *substance* foreign to the body that evokes an immune response"[115], but, also could be placed in row 4 the intersection of The Final Cause and The Efficient Cause because it is referring to a *type* of dendritic cell, but, could also be placed in row 3 the intersection of The Final Cause and The Final Cause because they are part of the *immune system*)

---

[113] "toxin." Eugene M. Mccarthy, Online Biology Dictionary.

[114] "langerhans cell." Eugene M. Mccarthy, Online Biology Dictionary.

[115] "antigen." Merriam-Webster Dictionary.

Trichome: "A filamentous outgrowth; especially: an epidermal hair structure on a plant. A major function of the trichome is thought to be in plant defense against insects. Chemicals produced in the glandular tip can deter feeding or the trichome can physically prevent the insect from reaching and feeding on the leaf."[116] (In row 2 the intersection of The Final Cause and The Material Cause because *chemicals* produced in the glandular tip can deter *feeding*, but, could also be placed in row 3 the intersection of The Final Cause and The Final Cause because it is thought to be a plant *defense* against insects)

Wall: "A high thick masonry structure forming a long rampart or an enclosure chiefly for defense —often used in plural."[117] (In row 2 the intersection of The Final Cause and The Material Cause because it is made out of *masonry*, but, also could be placed in row 5 the intersection of The Final Cause and The Formative Cause because it is a *high thick* masonry structure forming a *long* rampart or an *enclosure*, but, could also be placed in row 3 the intersection of The Final Cause and The Final Cause because it is chiefly for defense)

Cell wall: "In some eukaryotic cells, a rigid capsule enclosing the plasma membrane; in plants it contains cellulose and lignin; in fungi, chitin; in prokaryotes, a stiff capsule enclosing the cell membrane."[118] "The outermost layer of cells in plants, bacteria, fungi, and many algae that gives shape to the cell and protects

---

[116] "trichome." <https://quizlet.com>.

[117] "wall." Merriam-Webster Dictionary.

[118] "cell wall." Eugene M. Mccarthy, Online Biology Dictionary.

it from infection. In plants, the cell wall is made up mostly of cellulose. Most animal cells have a cell membrane rather than a cell wall."[119] (In row 2 the intersection of The Final Cause and The Material Cause because its composition and specific properties, but, could also be placed in row 3 the intersection of The Final Cause and The Final Cause because it is a wall for protection)

Lignin: "A hard material that joins with cellulose to form stiff cell walls in vascular plants; it also cements cells together providing structural strength to the plant as a whole."[120] (In row 2 the intersection of The Final Cause and The Material Cause because it is a *material* and is *hard*, but, could also be placed in row 3 the intersection of The Final Cause and The Final Cause because its purpose is to form stiff cell walls for *protection*)

Exudate: "An exudate is a fluid emitted by an organism through pores or a wound, a process known as exuding."[121] Animal and human exudates pertain to any fluid oozing out from the blood vessels, especially as a result of inflammation. When infection is present, the discharged fluid may contain white blood cells. In instances wherein there is vascular damage, red blood cells may also escape and be found in the exudate. Plant exudates include viscous materials seeping from interstices or pores. Examples of exudates include saps, gums, resins, and

---

[119] "cell wall." <https://www.thefreedictionary.com>.

[120] "lignin." Eugene M. Mccarthy, Online Biology Dictionary.

[121] "exudate." Wikipedia.

latex."[122] (In row 4 the intersection of The Final Cause and The Efficient Cause because it is the result of a *process* of inflammation, but, could also be placed in row 3 the intersection of The Final Cause and The Final Cause because sometimes has a protective function)

Reserve: "A supply of a commodity not needed for immediate use but available if required."[123] (In row 2 the intersection of The Final Cause and The Material Cause because it is a *commodity*, but, could also be placed in row 3 the intersection of The Final Cause and The Final Cause because is not needed for immediate use but available if *required*; in other words it is a stock of a resource)

Stockpile: "A large accumulated stock of goods or materials, especially one held in reserve for use at a time of shortage or other emergency."[124] (In row 2 the intersection of The Final Cause and The Material Cause, but, could also be placed in row 3 the intersection of The Final Cause and The Final Cause because is held in reserve for use at a time of *shortage or emergency*)

Cyst: "In an animal or plant, a thin-walled, hollow organ or cavity containing a liquid secretion; a sac, vesicle, or bladder. Medicine: in the body, a membranous sac or cavity of abnormal character containing fluid. ... A tough protective capsule enclosing the larva of a parasitic worm or the resting

---

[122] "exudate." Biology Online.

[123] "reserve." <https://quizlet.com>.

[124] "stockpile." <https://quizlet.com>.

stage of an organism."[125] (In row 3 the intersection of The Final Cause and The Final Cause because it stores potentially harmful elements)

Wax: "Waxes are similar to fats except that waxes are composed of only one long-chain fatty acid bonded to a long-chain alcohol group attached. ... Plants most noticeably use waxes for a thin protective covering of stems and leaves to prevent water loss. Similarly, animals employ waxes for protective purposes; for instance, earwax in humans prevents foreign material from entering and possibly injuring the ear canal area."[126] (In row 2 the intersection of The Final Cause and The Material Cause because wax is a compound, but, it also could be placed in row 3 the intersection of The Final Cause and The Final Cause because among the purposes of God's creation is our protection and that of animals and plants)

Of course, as in the case of the chair, each one of those words can be perceived from different perspectives. What is the potential of being able to classify those acceptations in this cause?

The same grouping criteria applies to the below mentioned faculties associated with the Final Cause.

OTHER HUMAN FACULTIES LINKED TO THE FINAL CAUSE

There is a set of human faculties that is common to all human beings and with it we respond to any menace, injury, danger or hostility. Besides our immune system, a survival instinct and reflexes, human beings

---

[125] "cyst." <https://quizlet.com>.

[126] "wax." <https://quizlet.com>.

have been endowed with some other faculties to protect ourselves and others:

Sense of Obligation also Called Sense of Responsibility

When considering "Why?" and "For what?" it is very important to recognize not just and individual but also a collective sense of responsibility.

'Abdu'l-Bahá said: "O ye lovers of truth, ye servants of humankind! Out of the flowering of your thoughts and hopes, fragrant emanations have come my way, wherefore an inner *sense of obligation* compelleth me to pen these words."[127]

Shoghi Effendi says: "... Nothing but a fiery ordeal, out of which humanity will emerge, chastened and prepared, can succeed in implanting that *sense of responsibility* which the leaders of a newborn age must arise to shoulder."[128]

If religious and natural laws are inherent to human beings, why is science held to be non-responsible?

Mikael Stenmark says:

Science should be non-responsible in the sense that scientist have not special accountability for the application of science. The ends to which scientific results are to be applied should be determined by society. Within, for instance, the framework of democracy, people are equally allowed to use the teachings and the findings of science to whatever vision of a good human life they endorse, be it a feminist, a Christian, or a Buddhist vision. On the other hand, scientists qua

---

[127] Selections, p. 246. Emphasis Added.

[128] The World Order of Bahá'u'lláh, p. 46. Emphasis Added.

scientists should themselves no take side in these debates. To take a stand on questions about the utility of science is not part of the scientist' task, nor does scientists have any special responsibility that goes beyond those they have as citizens. The only thing scientist should care about in their professional role is finding out the truth, or how nature works.[129]

Consider the following: Using the qualitative method of science, a scientist may discover that a compound is a good painkiller. What if, later, the medical community discovers that the painkiller is highly addictive? Do you agree that the role of the scientist is limited to finding out if the new compound is an effective painkiller?

The Senses of Fear and Shame

When considering "Why?" and "For what?" we should be aware of the warning signs of our senses of fear, shame and our survival instinct; because all of them have been given to us for our own protection and others'.

> The *fear of God* hath ever been a sure defense and a safe stronghold for all the peoples of the world. It is the chief cause of the protection of mankind, and the supreme instrument for its preservation. Indeed, there existeth in man a faculty which deterreth him from, and guardeth him against, whatever is unworthy and unseemly, and which is known as his *sense of shame*. This, however, is confined to but a few; all have not possessed, and do not possess, it. It is incumbent upon the kings and the spiritual leaders of the world to lay fast hold on religion,

---

[129] Mikael Stenmark qtd. in LeRon Shults (ed.), The Evolution of Rationality, p. 51.

inasmuch as through it the *fear of God* is instilled in all else but Him.[130]

However, I believe we can pray and ask God, the All Generous, to provide us with the sense of shame as a powerful instrument for our protection and that of others.

It is of the utmost importance to introduce the next argument setting as a foundation an absolute respect for the free will of the individual so he (she) can assume the outcome of his (her) decision:

Let us now briefly examine alcoholic drinks in the context of the above-mentioned set of human faculties to perceive its importance when addressing issues: Abdu'l-Bahá explains that the Aqdas prohibits "both light and strong drinks", and He states that the reason for prohibiting the use of alcoholic drinks is because "alcohol leadeth the mind astray and causeth the weakening of the body."[131] Further, Abdu'l-Bahá says: "Alcohol consumeth the mind and causeth man to commit acts of absurdity, but this opium, this foul fruit of the infernal tree, and this wicked hashish extinguish the mind, freeze the spirit, petrify the soul, waste the body and leave man frustrated and lost."[132]

Science supports this rationality, and research shows that alcohol is a depressant of the central nervous system, it is not a stimulant. It depresses our immune system[133] and our reflexes. Alcohol also

---

[130] Bahá'u'lláh, Epistle to the Son of the Wolf, p. 27. Emphasis Added.

[131] Bahá'u'lláh, The Kitáb-i-Aqdas, p. 227, Note 144.

[132] ibid. 239, Note 170.

[133] The National Institute on Alcohol Abuse and Alcoholism says: "A number of reviews in the literature provide an overview of current knowledge concerning alcohol's effects on the human immune system (Baker and Jerrells 1993; Cook 1995, 1998; Frank and Raicht 1985; Ishak et al. 1991; Johnson and Williams 1986;

compromises our intelligence and depresses our senses of responsibility, shame and of fear[134]. And of course, it depresses all human faculties. Yet the alcoholic beverages industry labels alcohol as causing euphoria; but besides being a false sense of wellbeing, only lasts while the individual consumes between one to four drinks and then disappears. How many chronic illnesses are caused by alcohol?[135] I leave to the reader to find out if alcohol disconnects the neuron's synapsis or kills the neurons. What are the consequences of having our memory affected?[136]

Kanagasundaram and Leevy 1981; MacGregor and Louria 1997; Mendenhall et al. 1984; Mufti et al. 1989; Palmer 1989; Paronetto 1993; Watson et al. 1986. Web. December 2016 < https://pubs.niaaa.nih.gov/publications/10report/chap04b.pdf >.

[134] There are many life histories of Alcoholics Anonymous members that can serve to develop our awareness of what can happen to our senses of shame, fear and responsibility. There are also many books that dial with this subject such as *The Treatment of Shame and Guilt in Alcoholism Counseling* By Ronald T. Potter-Efron, Patricia S. Potter-Efron and in *From Guilt Through Shame to AA: A Self-Reconciliation Process* by Ed Ramsey 1987 Alcoholism Treatment Quarterly 4(2) 87-107.

[135] To get a glimpse of the consequences of alcohol consumption there are several interesting studies by Jürgen Rehm who is the Director, Social and Epidemiological Research Department and Senior Scientist and Head of Group Population Health Research at the Centre for Addictions and Mental Health (CAMH) in Toronto. One of his publications is: Rehm, Jürgen, Jens Klotsche and Jayadeep Patra. *Comparative quantification of alcohol exposure as risk factor for global burden of disease.* International Journal of Methods in Psychiatric Research. June 2007. Volume 16, Issue 2 66–76. Web. January 2013

[136] "Korsakoff's syndrome (also called Korsakoff's dementia, Korsakoff's psychosis, or amnestic-confabulatory syndrome) is a neurological disorder caused by a lack of thiamine (vitamin B_1) in the brain or viral encephalitis. Its onset is linked to chronic

Suppose that in an act of mercy the administrative order of the town I live in, decided to carefully educate me and others about why alcoholic beverages are prohibited. However, one day I decided to drive a car under the influence of alcohol and I killed somebody. Do I not deserve a severe punishment? How can I asked for mercy? The institution should keep me in jail, in accordance with the law, showing its mercifulness to the public.

> The object of punishment is not vengeance, but the prevention of crime.
>
> Kings must rule with wisdom and justice; prince, peer and peasant alike have equal rights to just treatment, there must be no favour shown to individuals. A judge must be no "respecter of persons", but administer the law with strict impartiality in every case brought before him.
>
> If a person commits a crime against you, you have not the right to forgive him; but the law must punish him in order to prevent a repetition of that same crime by others, as the pain of the individual is unimportant beside the general welfare of the people.[137]

Memory

When considering "Why?" and "For what?" it is very beneficial to listen carefully and remember the advice of those with experience. Also, in our memory we store what we have learned, for example: after several times of trying to do something we develop the skill.

---

alcohol abuse or severe malnutrition, or both." Wikipedia. Web. March 2015 <http://en.wikipedia.org/wiki/Korsakoff%27s_syndrome>

[137] 'Abdu'l-Baha, Paris Talks, p. 152

... it is from *memory* that men derive their experience. For many recollections of the same thing perform the function of a single experience. Indeed, it is thought that experience is more or less similar to knowledge and skill, and that men acquire knowledge and skill through experience. ...

Now the circumstances in which skill arises are from the many cases of thinking in experience a single general assumption is formed in connection with similar things. For instance, to have the assumption that when Callias is ill with such and such a disease such and such medicine is appropriate and similarly for Socrates and for many others individually is a matter of *experience*. But the knowledge that for all such people ... when ill with such and such a disease, such and such a medicine is beneficial belongs to skill.[138]

'Abdu'l-Bahá taught:

There are five outward material powers in man which are the means of perception—that is, five powers whereby man perceives material things. They are sight, which perceives sensible forms; hearing, which perceives audible sounds; smell, which perceives odours; taste, which perceives edible things; and touch, which is distributed throughout the body and which perceives tactile realities. These five powers perceive external objects.

Man has likewise a number of spiritual powers: the power of imagination, which forms a mental image of things; thought, which reflects upon the realities of things; comprehension, which

---

[138] Aristotle. The Metaphysics. Emphasis Added.

understands these realities; *and memory, which retains whatever man has imagined, thought, and understood*.[139]

Memory is among the set of faculties related to the final cause because memory is a repository of what we comprehend. Memory is similar to other kinds of reservoirs or storing organs in the vegetable, animal and human kingdoms to preserve nutrients, natural defences or to isolate toxic substances acting as differential responses to needs and menaces.

If we were to claim that all these effects proceed from the powers of the animal nature and the physical senses, then we see plainly and clearly that, with regard to these powers, the animals are superior to man. For example, the sight of animals is much keener than that of man, their hearing is more acute, and likewise, their powers of smell and taste. Briefly, in the powers which man and animal share in common, the animal often has the advantage. *Take the power of memory: If you carry a pigeon from here to a faraway country, and there set it free, it will remember the way and return home. Take a dog from here to the heart of Asia, set it free, and it will return home without ever losing its way.* And so is it with the other powers, such as hearing, sight, smell, taste, and touch.[140]

Somewhere else we find:

Immunological *memory* is a hallmark of the adaptive immune system. However, the ability to remember and respond more robustly against a second encounter with the same pathogen has been described in organisms lacking T and B cells. Recently, NK cells have been shown

---

[139] Some Answered Questions, p. 56. Emphasis Added.

[140] 'Abdu'l-Bahá, Some Answered Questions, p. 48. Emphasis Added.

to mediate Ag-specific recall responses in several different model systems. Although NK cells do not rearrange the genes encoding their activating receptors, NK cells experience a selective education process during development, undergo a clonal-like expansion during virus infection, generate long-lived progeny (i.e., memory cells), and mediate more efficacious secondary responses against previously encountered pathogens—all characteristics previously ascribed only to T and B cells in mammals. This review describes past findings leading up to these new discoveries, summarizes the evidence for and characteristics of NK cell *memory*, and discusses the attempts and future challenges to identify these *long-lived memory* NK cell populations in humans.[141]

*

In conclusion, one of the phases of the cycle of scientific research is the explicative phase; and the notions and the set of human faculties associated with the final cause are ideal to be able to answer the questions Why? and What for?

**

In his search for a single science in its relation to its *mission* Aristotle wrote:

And the science which knows to what *end* each thing must be done is the most authoritative of the sciences, and more authoritative than any ancillary science; and this *end* is the good of that thing, and in general the *supreme good* in the whole of nature. Judged by all the tests we have mentioned, then, the name in question falls to the same

---

[141] Sun, Joseph C. et al. Emphasis Added.

science; this must be a science that the first principles and causes; for the good, i.e. the *end*, is one of the causes.[142]

Let us consider again what the wise Diotima taught Socrates about love in connection to the words where emphasis has been added:

> … So that if a virtuous soul have but a little comeliness, he will be content to love and tend him, and will search out and bring to the birth thoughts which may improve the young, until he is compelled to contemplate and see the beauty of *institutions and laws*, and to understand that the beauty of them *all* is of one family, and that personal beauty is a trifle … .[143]

---

[142] The Metaphysics, Emphasis added.

[143] Plato, Symposium. Emphasis added.

VII THE MATERIAL CAUSE AND THE QUALITATIVE METHOD OF SCIENCE

In order to introduce the "Material Cause," let us continue with the example of the chair. The answer to the question "*What is the chair made of?*" implies, necessarily, everything that is inherently part of the chair in addition to the raw materials used (wood, fabric, nails, glue and paint). In other words, we have to include: the knowledge of the chair's builder, the specific properties of the wood utilized in its construction (such as its hardness, resistance, permeability, susceptibility to attacks by mold, and so on), and the cultural values that the chair represents, including the harmony it must have with the cultural values of those who will use it. It is also necessary to consider, when assessing the value of the chair—and to realize how profound a recognition it is——-, the honesty of the chair maker, in choosing the materials for its construction, and the honesty of the community where the chair will be used. This is part of the true value of the chair. We must perceive that the chair is not just a pile of molecules, of different substances and compounds, but also an expression of God's generosity that allows us to procure our own comfort and welfare. Even before the chair came into existence, the materials and compounds that later went into the chair had the potential to become a chair and, latent within us, was the knowledge of how to transform them into a chair.

The essence of the Spirit, as the material cause, was the way to conceive the fitting together of the set of notions and the corresponding set of human faculties associated with the qualitative method of science. It will help us to answer the questions: "Who am I?"

PHILOSOPHICAL ARGUMENT

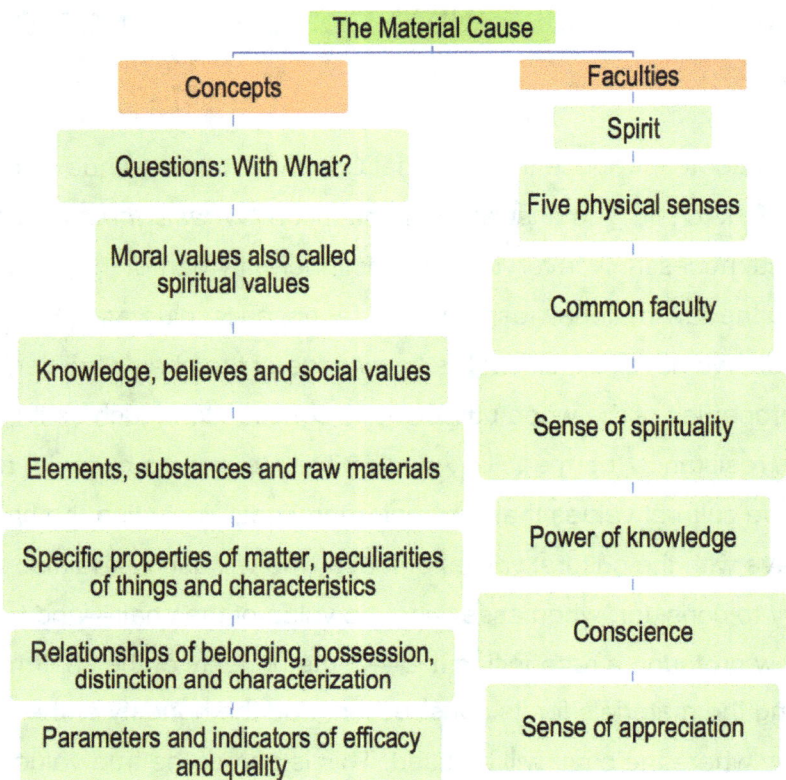

In *A Dictionary of Philosophical Terms and Names* we find "The Material Cause" defined as: "Basic stuff of which a thing is made" (Kemerling). In an English dictionary, the Material Cause is defined as something out of which something is made or comes into being (*Merriam Webster Dictionary*, 2012). This cause is associated with the question: *What is it made of?* or *With what?*

Ian Kluge has a very profound analysis of this topic:

Because God is 'beyond' the phenomenal realm, both the Baha'i Writings and Aristotle agree that God is *essentially unknowable* and do so for similar reasons. According to Aristotle, God, unlike all phenomena which are composed of *matter* and form, is one because the Divine has no *matter* and is pure form. The Divine is, moreover, pure existence, that is, a non-contingent entity whose nature is to exist; It is also pure thought thinking only on Itself. As time-and-space bound, *composite* beings, we can understand these concepts verbally, but cannot comprehend or understand what it is or means to enjoy this sort of being. Similarly, Abdu'l-Baha says,

> It is evident that the human understanding is a *quality* of the existence of man, and that man is a sign of God: how can the *quality* of the sign surround the creator of the sign?--that is to say, how can the understanding, which is a *quality* of the existence of man, comprehend God? Therefore, the *Reality* of the Divinity is hidden from all comprehension, and concealed from the minds of all men. It is absolutely impossible to ascend to that plane.[144]

By the "Reality of Divinity" Abdu'l-Baha means the *essence* of divinity which is beyond human comprehension. The *attributes* of divinity can, of course, be *known* or comprehended, but not the *essence* of Divinity. As pure form thinking Itself, Aristotle's God also enjoys a form of being whose nature can be deduced by Its *attributes* and actions in the phenomenal realm but cannot be *known* immediately. This is because, according to Aristotle, *true knowledge is knowledge* of causes and not mere description. That, however, is the level at which we must remain with the Unmoved Mover. [145]

---

[144] 'Abdu'l-Baha, Some Answered Questions, p. 146. Emphasis added.

[145] Kluge, Ian. The Aristotelian Substratum of the Baha'i Writings. Emphasis added.

From this we can conclude that the Baha'i Writings and Aristotle agree on several key epistemological issues subject to vociferous contemporary debate: first, that *natural reality* is *objectively real* and does not depend on human observers for its existence; second, that *reality* and its laws are given by God, not constructed, and that we must work with what is given; and third, that truth is the correspondence between *reality* and our interpretation of it, or, put otherwise, that *reality* and our interpretation of it are two distinct things and that we must test our interpretations against *reality* to discover whether or not they are in agreement. From this follows that reality is discovered and that there is such a thing as error, that is, an erroneous or inadequate understanding of reality that can be cured by abandoning it in order to change from *ignorant* to more *knowledgeable*. In other words, the Baha'i Writings and Aristotle share a *realist* epistemology. Without these premises, the entire Aristotelian and Baha'i enterprises would collapse, most especially the Baha'i doctrine of progressive revelation which presumes increasingly adequate comprehension of various truths. Finally, the *belief that properties are real* makes the Baha'i Writings and Aristotle incompatible with nominalism, that is, the *belief that properties* are either arbitrary human selections or outright impositions only externally related to their objects and that essences are fictitious. (See Aristotle's refutation of the underlying logic of nominalism in *Metaphysics*, VII, 12.) For its part, realism holds that the relationship between *attributes and substance* is internal, that is, inherent and intrinsic and that *essences are natural and real*.[146]

---

[146] Kluge, Ian. The Aristotelian Substratum of the Baha'i Writings. Emphasis added.

Analyzing the composition, the *What is it made of ...?,* of resources and services is the key to setting the priorities and objectives of individuals and institutions.

Are Spiritual Qualities Appreciated as Part of what a Chair is Made of?

The Bahá'í teachings are a source of wisdom to determine the composition of things; in other words, what is inherent to them:

> There are spiritual principles, or what some call human values, by which solutions can be found for every social problem. Any well-intentioned group can in a general sense devise practical solutions to its problems, but good intentions and practical knowledge are usually not enough. The essential merit of spiritual principles is that they not only present a perspective which harmonizes with that which is *inherent in human nature*, it also induces an attitude, a dynamic, a will, an aspiration, which facilitate the discovery and implementation of practical measures. Leaders of governments and all in authority would be well served in their efforts to solve problems if they would first seek to identify the principles involved and then be guided by them.[147]

The author decided to choose "the spirit" as the essential origination of the Material Cause, because 'Abdu'l-Bahá is reported to have said:

> According to the great ancient philosophers the words soul, mind and spirit implied the underlying principles of life; the essence was

---

[147] The Universal House of Justice, The Promise of World Peace, pp. 15-16. Emphasis added.

expressed under different names and these three terms designated the various functions of the absolute reality, or the operations of the one single essence; for instance, when they dealt with the sensations of emotion they called it the soul; when they desired to express that power which discovers the reality of phenomena they gave it the appellation of mind and *when they discussed the consciousness which pervades the world of creation they gave it the title of spirit.*[148]

It has also been said that the spirit, according to scientific philosophy, is made up of a simple, and therefore indivisible substance:

Therefore, it is evident that life is the expression of composition, and mortality, or death, is equivalent to decomposition. As the *spirit of man* is not composed of material elements, it is not subject to decomposition and, therefore, has no death. It is self-evident that the human spirit is simple, single and not composed in order that it may come to immortality, and it is a philosophical axiom that the individual or indivisible atom is indestructible.[149]

Elsewhere we find: "… knowledge is a human attribute but so is ignorance; truthfulness is a human attribute but so is falsehood; and the same holds true of trustworthiness and treachery, justice and tyranny, and so forth. In brief, every perfection and virtue, as well as every vice, is an attribute of man."[150]

If we make a chair valued for the hardwood that it is made of, our knowledge and honesty are reflected in the kind of wood that the chair is

---

[148] Divine Philosophy, p. 120. Emphasis added.

[149] 'Abdu'l-Bahá, The Promulgation of Universal Peace, p. 306. Emphasis added.

[150] 'Abdu'l-Bahá, Some Answered Questions, p. 64.

made of, and can be verified by a person knowledgeable of different types of wood and specifically of the type of wood that the chair supposedly is made out of.

Are Convictions and Cultural Values Also Appreciated as Part of What a Chair is Made of?

Hofstede expressed his perception of culture in the following terms:

> I treat culture as 'the collective programming of the mind which distinguishes the members of one group from another.' This is not a complete definition ..., but it covers what I have been able to measure. Culture, in this sense, includes systems of values: and *values are among the building blocks of culture.*
>
> *Culture* is to human collectivity what *personality* is to an individual. *Personality* has been defined by Guilford (1959) as 'the interactive aggregate of personal *characteristics* that influence the individual's response to its environment'. *Culture* could be defined as the interactive aggregated of *common characteristics* that influences a human group's response to its environment. *Culture* determines the *identity* of a human group in the same way as *personality* determines the *identity* of an individual.[151]

For instance, if the Guambiano indigenous community in Silvia, Colombia, were to stop using its traditional skirts and vests or to start buying raw materials from the Colombian textile industry, this would cause, in just one generation, the loss of the technological culture related to raising wool-producing animals. They would also lose the knowledge of how to

---

[151] Hofstede, Culture's Consequences, p. 21. Emphasis added.

spin, dye and weave the wool, the aggregate economic value of producing their own wool, leading ultimately to the loss of their cultural identity. Was this not exactly the argument made valiantly and peacefully by Mahatma Gandhi when the Indian nation decided not to purchase English cloth and other products made in England because it would be conducive to poverty? In truth, the interdependence of the regions of the world is necessary but this should not prevent us from finding ways to strengthen the local and regional economy by looking for more efficient technologies.

Fundamentals, convictions, cultural values, spiritual values, social and economic values and knowledge are words linked to the Material Cause.

Supplies

This category includes elements, substances, raw materials and manufacturing supplies that are incorporated materially in the product that is offered.

The Specific Properties of Matter

Above, we considered human spiritual qualities when trying to answer the question "With what are we going to achieve change?" Now, we will take into account the specific properties (color, hue, density, hardness, caloric capacity, temperature of fusion and ebullition, characteristic odor and taste, etc.) of material things such as minerals, plants, animals and human beings. The specific properties of matter are those that help us determine, for example, what materials we would choose in order to make a hammer that serves its purpose. For instance, the quality of steel required to make a hammer for a soft surface such as wood is different from that used for a hammer for a hardened substrate such as concrete.

Relationships

The following relationships help us determine the distinction that makes a thing or organism unique when comparing it to others. The Material Cause is associated with a set of important relationships such as ownership, possession, characterization, identification, distinction and many others. By using these relationships, we find out how the existence of each quality is specifically appreciated.

It is also important when deciding with what we are going to make something, to take into account the spiritual laws and the norms that rule quality, standard, possession or distinctiveness.

Parameters and Indicators

Of possession, identity, quality, efficacy and others can be used to monitor and evaluate the situation.

Applying the Single Science to the Material Cause

The table below shows the methodology followed in the construction of the "New Approach to Science" by inter-weaving the causes horizontally and vertically. In order for the reader to understand the placement of the following acceptations in one of the Rows (1 to 6), it is important to understand how the Material Cause is interwoven with the other causes.

	THE MATERIAL CAUSE: AND THE QUALITATIVE METHOD OF SCIENCE	
1	With what?	MATURITY CAUSE
2	Essence: The Spirit, Elements, substances and raw materials. Specific properties of matter, peculiarities of things and characteristics. Moral values, also called spiritual values. Knowledge, beliefs and social values	MATERIAL CAUSE
3	Spiritual laws, the standards that regulate quality and laws of possession **Goals** and objectives	FINAL CAUSE
4	**Powers:** conscience, knowledge, common faculty and senses of spirituality and appreciation **Specific lines of action or policies** to consolidate ethics, knowledge, quality and efficacy	EFFICIENT CAUSE
5	**Relationships** of belonging, possession, distinction and characterization	FORMATIVE CAUSE
6	**Parameters and indicators** of efficacy and quality in attaining the goals and objectives	MATURITY CAUSE

Having the basis of a conceptual framework for the Material Cause and learned in the single science proposed above in "Classes and Categories as foundation of Philosophy", that we are supposed to classify each notion of interest for the research in one of the causes; the following sample of a general accepted meaning of a word could be placed in the Material Cause:

"Truthfulness is the foundation of all the virtues of the world of humanity. Without truthfulness, progress and success in all of the worlds of God are impossible for a soul. When this holy attribute is established in man, all the divine qualities will also become realized."[152] (In row 2 the intersection of The Material Cause and The Material Cause because it is a *foundation* and is a *virtue*, but, it also could be placed in row 3 the intersection of The Material Cause and The Final Cause because it is a spiritual law: the inclusion of the following statements: "in all the worlds of God are impossible" and "all the divine qualities will also become realized" means the consequences of disobeying and obeying the law).

Prerequisite: "Something that you officially must have or do before you can have or do something else."[153] (In row 3 the intersection of The Material Cause and The Final Cause because you *must* have or *must* do before)

Cofactor: "Any molecule or ion required for an enzyme's function."[154] (In row 2 the intersection of The Material Cause and The Material Cause because it is a molecule or a ion; and, could also be placed in row 3 the intersection of The Material Cause and The Final Cause because it is *required* for an enzyme's function)

---

[152] Abdu'l-Baha, Tablets of Abdu'l-Baha v2, p. 459. Emphasis added.

[153] "prerequisite." Merriam-Webster Dictionary.

[154] "cofactor." Eugene M. Mccarthy, Online Biology Dictionary.

Coenzyme: "An organic molecule required for an enzyme's function. Most vitamins are coenzymes."[155] (In row 2 the intersection of The Material Cause and The Material Cause because it is a molecule; and, could also be placed in row 3 the intersection of The Material Cause and The Final Cause because it is *required* for an enzyme's function)

Vitamin: "Any of a wide variety of chemical substances required by the body's metabolism, but that cannot be synthesized by the body."[156] (In row 2 the intersection of The Material Cause and The Material Cause because it is a chemical substance, and, could also be placed in row 3 the intersection of The Material Cause and The Final Cause because it is *required* by the body's metabolism)

Calcium: "Silver-white metallic element. ... Vertebrates require relatively large amounts of calcium for the production and maintenance of bone. It is also essential to the function of nerves and muscles, and is a necessary cofactor for the enzymes involved in blood clotting and a variety of other bodily processes."[157] (In row 2 the intersection of The Material Cause and The Material Cause because it is a chemical element; and, could also be placed in row 3 the intersection of The Material Cause and The Final Cause because it is *required* for the production and

---

[155] "coenzyme." Eugene M. Mccarthy, Online Biology Dictionary.

[156] "vitamin." Eugene M. Mccarthy, Online Biology Dictionary.

[157] "calcium." Eugene M. Mccarthy, Online Biology Dictionary.

maintenance of bone, the function of nerves and muscles, blood clotting and a variety of other bodily processes)

Activation energy: "The amount of energy (E_a) required to convert a stable molecule into a reactive one. It is the energy needed to produce the unstable condition in which the energy state of the bonds of the reactants is raised to a level corresponding to the unstable transition state that precedes a chemical reaction."[158] (In row 3 the intersection of The Material Cause and The Final Cause because for the reaction to occur is *conditioned* by the amount of energy needed, but, it also could be placed in row 4 the intersection of The Material Cause and The Efficient Cause because *activation* energy is described as the *energy* needed to *produce* the unstable condition that precedes a chemical *reaction*, but, could also be placed in row 5 the intersection of The Material Cause and The Formative Cause because it is referring to *size* of the energy required)

Amino: "Relating to, being, or containing an amine group."[159] (In row 2 the intersection of The Material Cause and The Material Cause because it *contains* an amine group, but, also in row 4 the intersection of The Material Cause and The Efficient Cause because of relating to *being* or containing an amine *group*)

Knowledge: "Is a familiarity, awareness, or understanding of someone or something, such as facts, information, descriptions, or skills, which is acquired through experience or education by

---

[158] "activation energy." Eugene M. Mccarthy, Online Biology Dictionary.

[159] "amino." Eugene M. Mccarthy, Online Biology Dictionary.

perceiving, discovering, or learning."[160] (In row 2 the intersection of The Material Cause and The Material Cause because it becomes a possession *acquired* through experience or education by perceiving, discovering, or learning)

Of course, as in the case of the chair, each one of those words can be perceived from different perspectives. What is the potential of being able to classify those acceptations in this cause?

The same grouping criteria applies to the below mentioned faculties associated with the Material Cause.

OTHER HUMAN FACULTIES LINKED TO THE MATERIAL CAUSE

There is a set of faculties that is common to all human beings and with it we get an idea of the composition of things and appreciate the essential aspects of reality. Besides the power of spirit, human beings have been endowed with:

Senses

With our outer senses we appreciate the composition and perceive lack of quality, efficacy and similar situations and, of course, the specific properties of matter.

Common Faculty

The author initially associated the common faculty to the Material Cause by exclusion. Reflecting later about it, I realized that "Carbon", the chemical element, is part of a large number of compounds. We can apply

---

[160] "knowledge." Wikipedia.

this same idea to other chemical elements. The letter "a" is part of a considerable number of words.

In the following quote, 'Abdu'l-Bahá clearly explains the function of the common sense in relation with the inner and outer senses:

> There are five outward material powers in man which are the means of perception that is, five powers whereby man perceives material things. They are sight, which perceives sensible forms; hearing, which perceives audible sounds; smell, which perceives odours; taste, which perceives edible things; and touch, which is distributed throughout the body and which perceives tactile realities. These five powers perceive external objects.
>
> Man has likewise a number of spiritual powers: the power of imagination, which forms a mental image of things; thought, which reflects upon the realities of things; comprehension, which understands these realities; and memory, which retains whatever man has imagined, thought, and understood. *The intermediary between these five outward powers and the inward powers is a common faculty, a sense which mediates between them and which conveys to the inward powers whatever the outward powers have perceived. It is termed the common faculty as it is shared in common between the outward and inward powers.*
>
> For instance, sight, which is one of the outward powers, sees and perceives this flower and conveys this perception to the inward power of the common faculty; *the common faculty* transmits it to the power of imagination, which in turn conceives and forms this image and transmits it to the power of thought; the power of thought reflects upon it and, having apprehended its reality, conveys it to the power of comprehension; the comprehension, once it has understood it,

delivers the image of the sensible object to the memory, and the memory preserves it in its repository.[161]

Conscience

With the sensibilities of our conscience we appreciate cultural values and our knowledge or ignorance about the level of trustworthiness, confidence, generosity, courtesy, honesty and many other values in ourselves; and also we grasp the degree of perfection of the specific properties of our own health and wellbeing.

Three quotes suffice in this respect:

"Man is the discoverer of the mysteries of nature; nature is not *conscious* of those mysteries herself. It is evident, therefore, that man is dual in aspect: as an animal he is subject to nature, but in his spiritual or *conscious being* he transcends the world of material existence."[162]

> ... the *conscience* of man is sacred and to be respected; and that liberty thereof produces widening of ideas, amendment of morals, improvement of conduct, disclosure of the secrets of creation, and manifestation of the hidden verities of the contingent world. Moreover, if interrogation of conscience, which is one of the private possessions of the heart and the soul, take place in this world, what further recompense remains for man in the court of divine justice at the day of general resurrection? Convictions and ideas are within the scope of the comprehension of the King of kings, not of kings; and soul and

---

[161] Some Answered Questions, p. 56. Emphasis added

[162] 'Abdu'l-Bahá, The Promulgation of Universal Peace, p. 81. Emphasis added.

conscience are between the fingers of control of the Lord of hearts, not of [His] servants.[163]

"The principle of obedience to government does not place any Bahá'í under the obligation of *identifying* the teaching of His Faith with the *political* program enforced by the government. For such an *identification*, besides being erroneous and contrary to both the *spirit* as well as the form of the Bahá'í Message, would necessarily create a conflict within the *conscience* of every loyal believer."[164]

Sense of Spirituality

It is so fascinating to acknowledge that all human beings have been endowed with a sense of spirituality to serve as the foundation to improve our own quality of life and that of others.

If we want to attain the best of what we have been endowed with, we have to determine the best approach to strengthen our sense of spirituality. The guidance to reach such a goal is here:

In explaining mysticism, Shoghi Effendi says:

For the core of religious faith is that mystic feeling which unites Man with God. This state of spiritual communion can be brought about and maintained by means of meditation and prayer. And this is the reason why Bahá'u'lláh has so much stressed the importance of worship. It is not sufficient for a believer merely to accept and observe the

---

[163] 'Abdu'l-Bahá, A Traveller's Narrative, p. 91. Emphasis added.

[164] Compilations, Lights of Guidance, p. 446. Emphasis added.

teachings. He should, in addition, cultivate *the sense of spirituality* which he can acquire chiefly by means of prayer.[165]

Power of Knowledge

We determined the composition of an object with our knowledge of two things: first, the specific properties of matter, such as density, hardness and many others; and second, the moral and social values of the environment with our spiritual qualities.

Let me include two quotes from the Master:

"O my dear Mr. . . .! I beseech God to grant thee the power of knowledge and understanding so that thou mayest fathom the mysteries of the teachings descended from the presence of the Glorious Lord."[166]

"The Lord of the Kingdom hath invited, chosen and guided you through His pure favor, feeding you from the heavenly table of divine knowledge! Know ye the value of this favor and bounty and loosen your tongues in praise; showing forth the power of knowledge and assurance and breathing the spirit of guidance into the hearts of the seekers."[167]

Additionally, it is knowledge that distinguishes an individual from others. In other words, an individual's scientific, artistic, moral and technological culture can set that individual apart from others.

In order to clarify concepts such as culture, true wealth, true necessity, poverty, health, and wellbeing, the conceptual framework is used to inquire about which cultural and spiritual values—and what knowledge—are required to solve the issues to which the framework is to be applied. It

---

[165] Directives from the Guardian, p. 86. Emphasis added.

[166] 'Abdu'l-Bahá, Tablets of 'Abdu'l-Bahá, v2 p. 454.

[167] ibid. v2, p. 478.

is here where we have to apply the forces of the transformational process of society.

As a scientist I agree with Richard Feynman: "Science alone of all the subjects contains within itself the lesson of the danger of belief in the infallibility of the greatest teachers[168] in the preceeding generation . . .As a matter of fact, I can also define science another way: Science is the belief in the ignorance of experts."[169]

I also agree with Ridley:

> The fuel on which science runs is ignorance. Science is like a hungry furnace that must be fed logs from the forests of ignorance that surround us. In the process, the clearing that we call knowledge expands, but the more it expands, the longer its perimeter and the more ignorance comes into view. . . . A true scientist is bored by knowledge; it is the assault on ignorance that motivates him - the mysteries that previous discoveries have revealed. The forest is more interesting than the clearing.[170]

Sense of Appreciation

I believe that this is the sense to be polished and strengthened in order to perceive just the good in others and not their defects.

> With regard to your question concerning the meaning of the name 'Hidden Words'. It is, indeed, one of the most suggestive titles of the Writings of Bahá'u'lláh. These words are called hidden due to the fact that men have had *neither the knowledge nor a true sense of*

---

[168] The author assumes that Feynman is not referring to the Manifestations of God.

[169] The Pleasure of Finding Things Out.

[170] Genome: the autobiography of a species in 23 chapters.

appreciation of them before they were revealed by Bahá'u'lláh. It is through Him, Who is the sole Mouthpiece of God in this age, that spiritual realities and truths have been once more reinterpreted and revealed afresh to mankind. Bahá'u'lláh's Message is thus the only key to a true revealed afresh to mankind. Bahá'u'lláh's Message is thus the only key to a true understanding of the mysteries that envelop man's spiritual life.[171]

Shoghi Effendi, the beloved Guardian, said:

"I shall indeed grieve if the situation in Palestine should prevent our meeting and prevent your pilgrimage to the Holy Shrines. I pray that this may not be the case. I am so eager to meet you, and express in person my deep and abiding *sense of appreciation* of the splendid and historic services you have rendered. I will continue to pray for you from the depths of my heart."[172]

I accept Richard Feynman's statement "*Religion is a culture of faith; science is a culture of doubt*" and even Popper's "falsifiability"[173] for laws, principles, compositions, methods and cycles, when proposed by human beings. But, taking into account a sense of appreciation when referring to the Holy Scriptures, Richard Feynman's statement "*Religion is a culture of faith; science is a culture of doubt*" and Popper's "falsifiability should be change for "*Religion is a culture of faith; science is a culture of*

---

[171] From a letter written on behalf of Shoghi Effendi to an individual believer, 1 September 1935. Compilations, Lights of Guidance, p. 488. Emphasis added.

[172] Shoghi Effendi, Arohanui - Letters to New Zealand, p. 48. Emphasis added.

[173] The belief that for any hypothesis to have credence, it must be inherently disprovable before it can become accepted as a scientific hypothesis or theory. Web. March 2015 <https://explorable.com/falsifiability>.

reaffirmation", because science's role is to find the facts that explains the teachings of God's Messengers in the context of Progressive Revelation and Their influence in the evolving organizational structure of tribes into cities, then city-states, then nations, and the promise of a world federated government. For example: "In Moses time tribes where nomadic and there where not tribunals, so the law 'Eye for eye, and tooth for tooth' was the only way to be protected until they settled and changed this practice."[174]

*

This set of human faculties can help us answer the questions: "What am I?" and play a critical role when making ethical choices.

Another one of the phases of the cycle of scientific research is the qualitative phase, and the set of human faculties associated with the Material Cause are appropriate to answer the question: "With what?"

**

In his search for a single science Aristotle wrote:

> The first problem concerns the subject which we discussed in our remarks. It is this-(1) whether the investigation of the causes belongs to one or to more sciences, and (2) whether such a science should survey only the first principles of substance, or also the principles on which all men base their proofs, e.g. whether it is possible at the same time to assert and deny one and the same thing or not, and all other such questions; and (3) if the science in question deals with substance, whether one science deals with all substances, or more than one, and if more, whether all are akin, or some of them must be called forms of Wisdom and the others something else. And (4) this itself is also one of the things that must be discussed-whether

---

[174] Morgenstern, Julian. A Jewish Interpretation of the Book of Genesis, p. 74.

> sensible substances alone should be said to exist or others also besides them, and whether these others are of one kind or there are several classes of substances, as is supposed by those who believe both in Forms and in mathematical objects intermediate between these and sensible things. Into these questions, then, as we say, we must inquire, and also (5) whether our investigation is concerned only with substances or also with the essential attributes of substances.[175]

Further down he wrote:

> In general, do all substances fall under one science or under more than one? If the latter, to what sort of substance is the present science to be assigned?-On the other hand, it is not reasonable that one science should deal with all. For then there would be one demonstrative science dealing with all attributes. For ever demonstrative science investigates with regard to some subject its essential attributes, starting from the common beliefs. Therefore to investigate the essential attributes of one class of things, starting from one set of beliefs, is the business of one science. For the subject belongs to one science, and the premisses belong to one, whether to the same or to another; so that the attributes do so too, whether they are investigated by these sciences or by one compounded out of them.[176]

Let's appreciate again what the wise Diotima taught Socrates about her opinion of love in the context of reading again a quote about the quality

---

[175] The Metaphysics.

[176] ibid.

of love, focusing in connection to the words where emphasis has been added.

As an introduction to her statement let's read what The Master says

> It is impossible to consider this *life* apart from the future *life*. It is all one great whole. The thought of *what* is to come after death is not only a great comfort in times of earthly stress and suffering, but is also a powerful influence toward *right conduct in this life*.
>
> 'Abdu'l-Bahá has said that without this vision of the next *life* there cannot be enough incentive to *ethical* action here.[177]

Diotima said:

> 'He who has been instructed thus far in the things of love, and who has learned to see the beautiful in due order and succession, when he comes toward the end will suddenly perceive a *nature* of wondrous beauty (and this, Socrates, is the final cause of all our former toils)-a *nature* which in the first place is everlasting, not growing and decaying, or waxing and waning; secondly, not fair in one point of view and foul in another, or at one time or in one relation or at one place fair, at another time or in another relation or at another place foul, as if fair to some and-foul to others, or in the likeness of a face or hands or any other part of the bodily frame, or in any form of speech or *knowledge*, or existing in any other being, as for example, in an animal, or in heaven or in earth, or in any other place; but beauty absolute, separate, *simple*, and everlasting, which without diminution and without increase, or any change, is imparted to the ever-growing and perishing beauties of all other things. He who from these

---

[177] SOW - Star of the West, Star of the West - 8. Emphasis added.

ascending under the influence of true love, begins to perceive that beauty, is not far from the end. And the true order of going, or being led by another, to the things of love, is to begin from the beauties of earth and mount upwards for the sake of that other beauty, using these as steps only, and from one going on to two, and from two to all fair forms, and from fair forms to fair practices, and from fair practices to fair notions, until from fair notions he arrives at the notion of absolute beauty, and at last *knows* what the *essence* of beauty is.

This, my dear Socrates,' said the stranger of Mantineia, 'is that *life* above all others which man should *live*, in the contemplation of beauty absolute; a beauty which if you once beheld, you would see not to be after the measure of *gold*, and *garments*, and fair boys and youths, whose presence now entrances you; and you and many a one would be content to *live* seeing them only and conversing with them without *meat* or *drink*, if that were possible-you only want to look at them and to be with them. But what if man had eyes to see the true beauty-the divine beauty, I mean, *pure and dear and unalloyed, not clogged with the pollutions of mortality and all the colours and vanities of human life*-thither looking, and holding converse with the *true beauty simple and divine*? Remember how in that communion only, beholding beauty with the eye of the mind, he will be enabled to bring forth, not images of beauty, but *realities* (for he has hold not of an image *but of a reality*), and bringing forth and *nourishing true virtue* to become the friend of God and be *immortal*, if *mortal* man may. Would that be an *ignoble life*?'[178]

---

[178] Plato, Symposium. Emphasis added.

VIII THE EFFICIENT CAUSE AND THE EXPERIMENTAL METHOD OF SCIENCE

In order to introduce the "Efficient Cause" let us continue with the example of the chair. In order to comprehend, "*Who made the chair*", let us think of the capacities of the carpenter and the methods he employed. We should also take into account the tools and equipment the carpenter used to make it, including those materials that are not directly incorporated in the chair, such as the sandpaper used to polish it, the rag that was used to dust it before it was painted, and so on. But much more profound is to admit that within the natural essence of those substances and compounds needed to make a chair, the potential to develop arts and tools in order to transform them into a chair, existed in a latent stage, even before any chair was made.

The essence of the Word of God, as the efficient cause, was the key used by the author to generate the set of notions and the corresponding set of human faculties associated with the Descriptive and Experimental methods of science. It will also help us answer the questions: "Who am I?" and we will be able to direct our efforts to address our misconceptions and wrongdoings.

PHILOSOPHICAL ARGUMENT

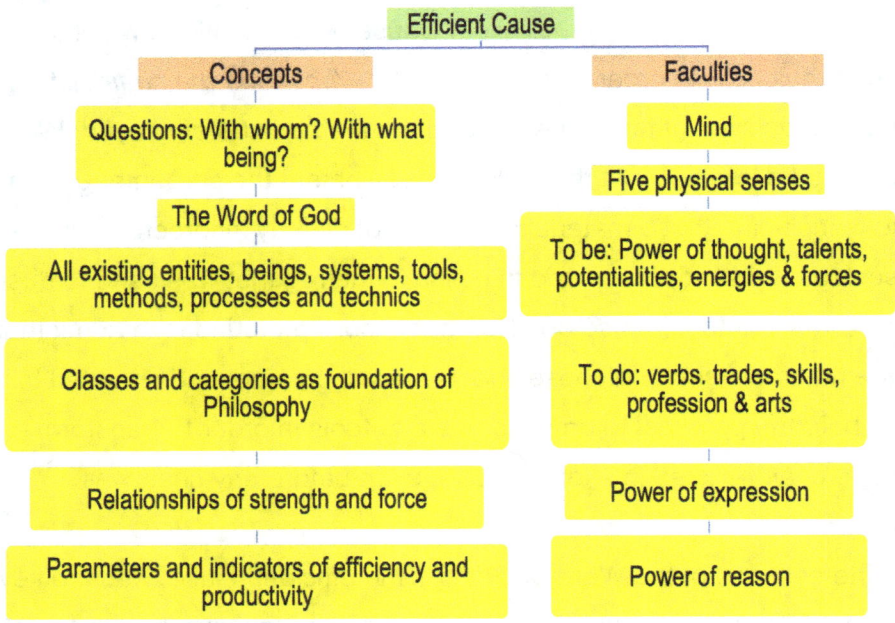

The Efficient Cause is defined as the true principle of change. In other words, it is the cause that, through its action, produces an entity called "effect". The Efficient Cause is associated with the question *"With whom?"* or "With what being?" It is associated with the faculties of doing, teaching and learning. We also find it associated with the concept of category, type or set; in other words, with the capacity to classify notions into groups.

The Efficient Cause is associated with concepts such as movement and the verbs "to be" and "to do". In Plato's Dialogues and in The Sophist, it is obvious that these are the fundamental aspects of the discussion between Theaetetus and the Stranger:

Stranger: Let us push the question; for if they will admit that any, even the smallest particle of being, is incorporeal, it is enough; they must then say what that nature is which is common to both the corporeal and incorporeal, and which they have in their mind's eye when they say of both of them that they 'are.' Perhaps they may be in a difficulty; and if this is the case, there is a possibility that they may accept a notion of ours respecting the nature of being, having nothing of their own to offer.

Theaetetus: What is the notion? Tell me, and we shall soon see.

Stranger: My notion would be, that anything which possesses any sort of power to affect another, or to be affected by another, if only for a single moment, however trifling the cause and however slight the effect, has real existence; and I hold that the definition of being is simply power.

Theaetetus: They accept your suggestion, having nothing better of their own to offer.[179]

The Mind

"As for the mind, it is the power of the human spirit. The spirit is as the lamp, and the mind as the light that shines from it. The spirit is as the tree, and the mind as the fruit. The mind is the perfection of the spirit and a necessary attribute thereof, even as the rays of the sun are an essential requirement of the sun itself."[180]

'Abdu'l-Bahá is reported to have said:

---

[179] Plato, Sophist.

[180] 'Abdu'l-Bahá, Some Answered Questions, p. 55.

There is, however, a faculty in man which unfolds to his vision the secrets of existence. It gives him a power whereby he may investigate the reality of every object. It leads man on and on to the luminous station of divine sublimity and frees him from all the fetters of self, causing him to ascend to the pure heaven of sanctity. This is the power of the mind, for the soul is not, of itself, capable of unrolling the mysteries of phenomena; *but the mind can accomplish this and therefore it is a power superior to the soul.*[181] And: "... Here reason signifieth the divine, universal mind, whose sovereignty enlighteneth all created things."[182]

"This supreme emblem of God stands first in the order of creation and first in rank, taking precedence over all created things. Witness to it is the Holy Tradition, "Before all else, God created the mind." From the dawn of creation, it was made to be revealed in the temple of man."[183]

"To be" and "To do"

The efficiency of an individual, a group, an institution or an enterprise is the result of the relationship between the power of the mind, the potentialities of the individuals, the equipment and materials available; and the methodologies, techniques, skills, procedures, habits and practices.

Think of beings as any noun, thing, entity, system, animate or inanimate object, creature, plant, animal or human being when being asked: *Who? What tool?* Or, even better, to include a question such as: *What is the agent that is causing this effect?* Here we include all supplies, tools and

---

[181] Divine Philosophy, p. 121. Emphasis added.

[182] Bahá'u'lláh, The Seven Valleys and the Four Valleys, p. 52.

[183] 'Abdu'l-Bahá, The Secret of Divine Civilization, p. 1. Emphasis added.

equipment that may help in the process to reap a product or a service and also fruits, byproducts and waste.

When thinking about to 'do', we should include movements, verbs, methods, processes, activities, procedures, lines of actions, techniques, skills, abilities, arts, technologies, and mechanisms. In order to be able of doing something we have to include concepts such as energy, forces, capacities, potentialities, talents, vocations, arguments and concepts.

> Know that nothing which exists remains in a state of repose -- that is to say, all things are in motion. Everything is either growing or declining; all things are either coming from nonexistence into being, or going from existence into nonexistence. So this flower, this hyacinth, during a certain period of time was coming from the world of nonexistence into being, and now it is going from being into nonexistence. This state of motion is said to be essential -- that is, natural; it cannot be separated from beings because it is their essential requirement, as it is the essential requirement of fire to burn.[184]

The Word of God

The Efficient Cause implies an active agent, which is external to the object, but acts upon it and generates an effect. We can associate this cause with other concepts such as the forces in nature, energy, heat, work, word, discourse, the Divine argument of the authorized interpreters of the Word of God, and human argument. In the words of Plato:

> **Stranger-** But now that the imitative art has enclosed him, it is clear that we must begin by dividing the art of creation; for imitation is a

---

[184] Abdu'l-Baha, Some Answered Questions, p. 233.

kind of creation of images, however, as we affirm, and not of real things.

Theaetetus- Quite true.

Stranger- In the first place, there are two kinds of creation.

Theaetetus- What are they?

Stranger- One of them is human and the other divine.

Theaetetus- I do not follow.

Stranger- Every power, as you may remember our saying originally, which causes things to exist, not previously existing, was defined by us as creative.

Theaetetus- I remember.

Stranger-Looking, now, at the world and all the animals and plants, at things which grow upon the earth from seeds and roots, as well as at inanimate substances which are formed within the earth, fusile or non-fusile, shall we say that they come into existence-not having existed previously-by the creation of God, or shall we agree with vulgar opinion about them?[185]

Continuing with Plato:

"**Stranger**- When any one says "A man learns," should you not call this the simplest and least of sentences?

Theaetetus- Yes."[186]

Bahá'u'lláh is even clearer and more emphatic, expressing:

---

[185] Plato, Theaetetus.

[186] ibid.

Know thou, moreover, that the Word of God -- exalted be His glory -- is higher and far superior to that which the senses can perceive, for it is sanctified from any property or substance. It transcendeth the limitations of known elements and is exalted above all the essential and recognized substances. *It became manifest without any syllable or sound and is none but the Command of God which pervadeth all created things. It hath never been withheld from the world of being.* It is God's all-pervasive grace, from which all grace doth emanate. It is an entity far removed above all that hath *been* and shall *be.*[187]

The author decided to choose "the Word of God" as the essential origination of the Efficient Cause, because Bahá'u'lláh said: *It hath never been withheld from the world of being.*

The Arts

The philosopher Socrates considered the purpose of the arts to be to complement science, a theme we will address below. In Theaetetus, or Of Science, Socrates distinguishes the arts from the sciences by saying that art is the object of science. He clarifies by saying:

> **Theaetetus**- Then, I think that the sciences which I learn from Theodorus-geometry, and those which you just now mentioned-are knowledge; and I would include the art of the cobbler and other craftsmen; these, each and all of, them, are knowledge.
>
> **Socrates**- Too much, Theaetetus, too much; the nobility and liberality of your nature make you give many and diverse things, when I am

---

[187] Tablets of Bahá'u'lláh, pp. 140 – 141. Emphasis added.

asking for one simple thing. **Theaetetus**- What do you mean, Socrates?

Socrates- Perhaps nothing. I will endeavour, however, to explain what I believe to be my meaning: When you speak of cobbling, you mean the art or science of making shoes?

Theaetetus- Just so.

Socrates- And when you speak of carpentering, you mean the art of making wooden implements?

Theaetetus- I do.

Socrates- In both cases you define the subject matter of each of the two arts? **Theaetetus**- True.

Socrates- But that, Theaetetus, was not the point of my question: we wanted to know not the subjects, nor yet the number of the arts or sciences, for we were not going to count them, but we wanted to know the nature of knowledge in the abstract. Am I not right?

Theaetetus- Perfectly right.[188]

There are many arts other than those that are traditionally recognized as such. In a passage that seemingly alludes to comedy and caricature, Socrates says:

Stranger- And is there any more artistic or graceful form of jest than imitation?

Stranger- We divided image-making into two sorts; the one likeness-making, the other imaginative or phantastic.

Theaetetus- True.

---

[188] Plato, Theaetetus.

Stranger- And may we not fairly call the sort of art, which produces an appearance and not an image, phantastic art?

Theaetetus- Most fairly.

Stranger- These then are the two kinds of image making-the art of making likenesses, and phantastic or the art of making appearances?[189]

In other paragraph we find a reference to diverse arts:

Stranger- Let us grant, then, that from the discerning art comes purification, and from purification let there be separated off a part which is concerned with the soul; of this mental purification instruction is a portion, and of instruction education, and of education, that refutation of vain conceit which has been discovered in the present argument; and let this be called by you and me the nobly-descended art of Sophistry.[190]

Perceiving the diverse perspectives from which reality can be seen, leads one to realize how difficult the art of consultation could become.

Art, then, is the result of the abilities, skills and capabilities of the individual and the potentialities of the other means employed to reap a fruit.

Relationships

Relations of strength and force are those that help us choose the most meaningful words in order to be able to describe a situation properly or to persuade others with reasonable arguments. The relationships

---

[189] Plato, Sophist.

[190] ibid.

between the potential capacity of the agent and the method and tools used to manifest it. Also, the relationships of all social actors with the labor force.

Parameters of Efficiency

For example: the efficiency of conversion of nutrients, rotation of inventories, rotation of accounts receivables, profitability, productivity, etc.

Applying the Single Science to the Efficient Cause

The table below shows the methodology followed in the construction of the "New Approach to Science" by inter-weaving the causes horizontally and vertically. In order for the reader to understand the placement of the following acceptations in one of the Rows (1 to 6), it is important to understand how the Efficient Cause is interwoven with the other causes.

	THE EFFICIENT CAUSE AND THE DESCRIPTIVE AND EXPERIMENTAL METHODS OF SCIENCE	
1	**Questions:** With what being? With whom? With what type?	MATURITY CAUSE
2	**Essence**: the Word of God. **To be:** animated and inanimated beings. **Forces:** power of mind, energy, capacities, capabilities, potential, talents, vocations, arguments, concepts. **Categories:** genres, clusters, groups. Interacting entities within a system or a subsystem.	MATERIAL CAUSE
3	**Laws** of thermodynamics, labor **law** and tools **regulations**. Grammar **rules**. The Word of God is Law **Fruits of labor:** results, harvest, services, products and leftover materials or efforts	FINAL CAUSE
4	**Powers:** thought, reason and expression **To do:** movements: methods, processes, activities, abilities, skills, arts, technologies, mechanisms.	EFFICIENT CAUSE
5	**Relations of**: cause and effect, logic and reason.	FORMATIVE CAUSE
6	**Parameters and indicators** of efficiency and productivity in achieving the results	MATURITY CAUSE

Having the basis of a conceptual framework for the Efficient Cause and learned in the single science proposed above in "Classes and Categories as foundation of Philosophy", that we are supposed to group each notion of interest for the solution of a problem in one of the causes, I believe that the following acceptations could be placed in the Efficient Cause:

System: "A regularly interacting or interdependent group of items forming a unified whole. A group of related parts that move or work together. A group of organs that work together to perform an important function of the body."[191] (In row 2 the intersection of The Efficient Cause and The Material Cause because the group is composed of *items*, *parts* or *organs*, but, it also could be placed in row 4 the intersection of The Efficient Cause and The Efficient Cause because is a *group* of *items* that *move* or *work* together)

Humoral immune system: "The portion of the immune system that produces antibodies that circulates in the blood and lymph."[192] (In row 3 the intersection of The Efficient Cause and The Final Cause because it is referring to the portion of the *immune system* that produces *antibodies*, but, could also be placed in row 2 the intersection of The Efficient Cause and The Material Cause because it is a *portion* of the immune system)

Photosynthesis: "A process carried out in plants, algae, and bacteria, which uses energy from sunlight to convert carbon dioxide and water into glucose and oxygen. Photosynthesis is the source of atmospheric free oxygen and is the essential starting point for the construction of all organic molecules present in living

---

[191] "system." Merriam-Webster Dictionary.

[192] "humoral immune system." Eugene M. Mccarthy, Online Biology Dictionary.

things."[193] (In row 4 the intersection of The Efficient Cause and The Efficient Cause because is a *process;* and, could also be placed in row 3 the intersection of The Efficient Cause and The Material Cause because Photosynthesis is the *essential* starting point for the construction of all *organic molecules* present in *living things*)

Altruism: "Unselfish behavior[194]; within a biological context, behavior that assists others to survive and reproduce, but that does not benefit the individual engaging in the behavior."[195] (In row 5 the intersection of The Efficient Cause and The Formative Cause because behavior refers to the way an animal or a person acts in *relation to others*, and, being unselfish is a *form of relationship* that influences the results. An unselfish behavior of an insect, bird or bat can't be considered a conscious decision of the individual, but, could also be placed in row 3 the intersection of The Efficient Cause and The Final Cause because the result of the altruistic behavior is to assists others to *survive* and reproduce)

Of course, as in the case of the chair, each one of those words can be perceived from different perspectives. What is the potential of being able to classify those acceptations in this cause?

The same grouping criteria applies to the below mentioned faculties associated with the Efficient Cause.

---

[193] "photosynthesis." Eugene M. Mccarthy, Online Biology Dictionary.

[194] https://plato.stanford.edu/entries/altruism-biological/

[195] "altruism." Eugene M. Mccarthy, Online Biology Dictionary.

OTHER HUMAN FACULTIES LINKED TO THE EFFICIENT CAUSE

There are other human faculties that are common to all human beings and with them we get an idea of the actors that transform reality.

Power of Thought

We can perceive ourselves as mere spectators of a reality built by others, or feel that we have the power to construct a better world.

"The reality of man is his *thought*, not his material body. The thought force and the animal force are partners. Although man is part of the animal creation, he possesses a power of thought superior to all other created beings."[196]

Let's have a reasonable exemplification to understand that the "reality of man is his thought not his material body" in a very interesting reflection made by William Huitt:

> Countries around the world are seeking ways to better prepare their children and youth for successful adulthood in the twenty-first century (Smith & Day, 1990; Jakobi & Teltemann, 2011). Unfortunately, there is no easy, readily available solution to this challenge. One reason is that the alternatives that one considers depends on the worldview and/or paradigm one uses to describe a human being and the value of education (Huitt, 2015). Therefore, discussion of alternatives is often an implicit discussion of worldviews and paradigms. For example, if one adopts a secular/materialistic worldview, one sees a human being in strictly materialistic terms whereas if one adopts a cosmic-spiritual worldview one would propose that there is some part of the human being that survives physical death. And if one were to

---

[196] 'Abdu'l-Bahá, Paris Talks, p. 24. Emphasis added.

adopt a God-centered worldview one would likely look to a set of scriptures or traditions to define a human being and a life after an earthly existence. Holders of each of these alternative worldviews will develop somewhat different alternatives and establish different criteria for choosing among them.

Likewise, one's paradigm can influence how one interprets reality and organizes facts, concepts, and principles (Huitt, 2011b). For example, if one were to adopt a mechanistic or reductionistic paradigm, one would look at formal education as a separate entity and investigate how the structure and functions of teachers and schools would impact human development (eg, Hattie, 2009; Squires, Huitt, & Segars, 1982). However, if one adopted an existential/phenomenological paradigm one would look at human perceptions and interpretations of schooling (Rogers, & Freiberg, 1994). And if one were to adopt an organismic/systems paradigm one would seek to describe the whole person embedded in multiple layers of context or environment (Huitt, 2012).[197]

"Man has likewise a number of spiritual powers: the power of imagination, which forms a mental image of things; *thought, which reflects upon the realities of things*; comprehension, which understands these realities; and memory, which retains whatever man has imagined, thought, and understood."[198]

---

[197] What Is A Human Being And Why Is Education Necessary.

[198] 'Abdu'l-Bahá, Some Answered Questions , p. 56. Emphasis added.

Power of Reason

The cause and effect relationships illumined by our reason lead us to understand the description of the results as a logical outcome of the chosen agent, its force and process, for example: If water is heated in a gas or electric stove, the molecules start moving quicker until the water boils.

Consider the following quotes:

> When we consider the third criterion—traditions—upheld by theologians as the avenue and standard of knowledge, we find this source equally unreliable and unworthy of dependence. For religious traditions are the report and record of understanding and interpretation of the Book. By what means has this understanding, this interpretation been reached? By the analysis of human reason. When we read the Book of God the faculty of comprehension by which we form conclusion is reason. Reason is *mind*. If we are not endowed with perfect *reason*, how can we comprehend the *meanings of the Word of God*? Therefore human reason, as already pointed out, is by its very nature finite and faulty in conclusions. It cannot surround the Reality Itself, the Infinite Word. Inasmuch as the source of traditions and interpretations is human reason, and human reason is faulty, how can we depend upon its findings for real knowledge?[199]

> In the vegetable world, too, there is the power of growth, and that power of growth is the spirit. In the animal world there is the sense of feeling, but in the human world there is an all-embracing power. In all the preceding stages the power of reason is absent, but the soul existeth and revealeth itself. The sense of feeling understandeth not

---

[199] 'Abdu'l-Bahá, Foundations of World Unity, p. 48. Emphasis added.

the soul, whereas the *reasoning power of the mind* proveth the existence thereof.[200]

Bahá'u'lláh has declared that religion must be in accord with science and reason. If it does not correspond with scientific principles and the processes of reason, it is superstition. For God has endowed us with faculties by which we may comprehend the realities of things, contemplate reality itself. If religion is opposed to reason and science, faith is impossible; and when faith and confidence in the divine religion are not manifest in the heart, there can be no spiritual attainment.[201]

Power of Expression

The capacity to express our thoughts and the results of our reasoning through our power of expression in a meaningful conversation is key for the transformation of the world into a better place. It can be achieved if the training of that power of expression is accompanied by developing the capacity to classify ideas.

Shoghi Effendi, the beloved Guardian, said:

> I feel the necessity of entrusting this highly important and delicate task to a special committee, to be appointed most carefully by the National Spiritual Assembly of America, and consisting of those who by their knowledge of the Cause, their experience in matters of publicity, and particularly by their power of expression and beauty of style will be

---

[200] 'Abdu'l-Bahá, 'Abdu'l-Bahá's Tablet to Dr. Forel, p. 2. Emphasis added.

[201] 'Abdu'l-Bahá, The Promulgation of Universal Peace, p. 1063.

qualified to produce a befitting statement on the unique history of the Movement as well as its lofty principles.[202]

*

This set of human faculties can help us to answer the question: "Who am I?"

In conclusion, the cycle of scientific research has also an experimental and/or descriptive phase and the notions and the set of human faculties associated with the Efficient Cause are appropriate to be able to answer the questions: "With whom?" and "What is the agent that is causing this effect?"

Thus, it is clear that the Efficient Cause can be linked to the concept of development, not defined as industrialization, but as the set of processes which aim to make manifest the latent potentialities of human beings and enable them to discover the potential of other beings in nature.

> Science is the most subversive thing that has ever been devised by man. It is a discipline in which the rules of the game require the undermining of that which already exists, in the sense that new knowledge always necessarily crowds out inferior antecedent knowledge. . . . . This is what the patent system is all about. We reward a man for subverting and undermining that which is already known. . . . . Man has a tendency to resist changing his mind. The history of the physical sciences is replete with episode after episode in which the discoveries of science, subversive as they were because they undermined existing knowledge, had a hard time achieving acceptability and respectability. Galileo was forced to recant; Bruno was burned at the stake; and so forth. An interesting thing about the

---

[202] Bahá'í Administration, p. 58.

physical sciences is that they did achieve acceptance. Certainly in the more economically advanced areas of the Western World, it has become commonplace to do everything possible to accelerate the undermining of existent knowledge about the physical world. The underdeveloped areas of the world today still live in a pre-Newtonian universe. They are still resistant to anything subversive, anything requiring change; resistant even to the ideas that would change their basic concepts of the physical world.[203]

Hauser wrote the above-mentioned ideas on the 20th century, but I believe that today's economically advanced areas of the Western World and also the underdeveloped areas of the world (a category that needs to be revised), will welcome the change in their basic concepts of the physical world proposed here, because the current conception of science is not responding to the true nature of what is a human being. Are we just mammals? Are we just insatiable consumers of material possessions and unappeasable in egotistical pleasures? Are we just sinners? Is it reasonable to fight endless wars to dominate others in order to impose a set of economic values or an ideology? For a profound reflection and very enriching discussion about this theme I want to refer the reader to *The Bahá'í Philosophy of Human Nature* by Ian Kluge[204].

Some intellectuals propose an interconnection between science and religion. I believe that our understanding of the Book of Creation greatly advances as we try to develop, in an attitude of respect, the right tools to interpret correctly the Book of Revelation[205]; and vice versa.

---

[203] Hauser, Philip M. Demographer and Census Expert, qtd. in Theodore Berland's The Scientific Life.

[204] The Journal of Bahá'í Studies 27.1-2 2017.

[205] The compilation of all the Holy Books.

When asked by an astounded atheist, if he were in fact deeply religious, Einstein replied: Yes, you can call it that. Try and penetrate with our limited means the secrets of nature and you will find that, behind all the discernible concatenations, there remains something subtle, intangible and inexplicable. Veneration for this force beyond anything that we can comprehend is my religion. To that extent I am, in point of fact, religious.[206]

However, we find the assertion of the celebrated physicist Stephen Hawking. When ABC News' Diane Sawyer asked if there was a way to reconcile religion and science, Hawking said, "There is a fundamental difference between religion, which is based on authority, [and] science, which is based on observation and reason. Science will win because it works."[207] In my opinion God's authority is explained when we humbly find out the scientific reasons of how wise He is.

Once the reader understands and tries this conceptual framework, he (she) will perceive that there is no conflict between science and religion. As we try and persevere to see them as in a peaceful coexistence, we will be able to establish peace in the also called blue planet when seen from space.

In The Proclamation of Bahá'u'lláh we find: "In such a world society, science and religion, the two most potent forces in human life, will be reconciled, will co-operate, and will harmoniously develop."[208] As the title suggest, when a new paradigm emerges by merging key notions from the realms of science and religion into one conceptual framework to perceive,

---

[206] Max Jammer, Einstein and Religion, pp. 39-40.

[207] Stephen Hawking qtd. in interview by ABC News' Diane Sawyer.

[208] Shoghi Effendi in the Introduction of a book by Bahá'u'lláh, The Proclamation of Bahá'u'lláh, p. XI.

interpret and transform reality, it becomes an extraordinary coalescence force to bring about the ideals of a new civilization.

****

In his search for a single science Aristotle wrote: "Now for each one *class* of things, as there is one perception, so there is one science, as for instance grammar, being one science, investigates all *articulate* sounds. Hence to investigate all the *species of being qua being*[209] is the work of a science which is generically one, and to investigate the several *species* is the work of the specific parts of the science."[210]

Let us read again what Diotima taught Socrates about the meaning of love in connection to the words where emphasis has been added: "… in the next stage he will consider that the beauty of *the mind is* more honourable than the beauty of the outward form. So that if a virtuous soul has but a little comeliness, he will be content to love and tend him, and will search out and bring to the birth *thoughts* which may improve the young, … ."[211]

---

[209] *Metaphysics* … is one of the principal works of Aristotle and the first major work of the branch of philosophy with the same name. The principal subject is "being qua being," or being insofar as it is being. It examines what can be asserted about anything that exists just because of its existence and not because of any special qualities it has. Also covered are different kinds of causation, form and matter, the existence of mathematical objects, and a prime-mover God.Web. January 2017 < https://en.wikipedia.org/wiki/Metaphysics_(Aristotle)>.

[210] The Metaphysics. Emphasis added.

[211] Plato, Symposium. Emphasis added.

IX THE MATURITY CAUSE AND THE PROPOSITIONAL METHOD OF SCIENCE

In order to introduce the "Maturity Cause in its relationship to the propositional method" let us advance with the example of the chair.

Consider now the options that we have after evaluating if a chair can be repaired. If the option is to repair it, then we may have evaluated the remaining expected useful life of the chair. If the decision is to trash it we may consider recycling its parts; if this is the option, then which parts should be recycled naturally for example by composting, and which parts can be reused in the industrial process?

But what if the option is to understand what happened that several chairs are broken, so we do not make the same mistakes when building new ones? Then, we should look into all the decisions made during each of the phases of the cycle(s) that could be related to the specific damage of the chair.

If the legs, made out of special wood, are showing recurrent damages, then we could go as far as finding out what kind of parameters and indicators were being used to monitor and evaluate what happens in the cycle of growing, harvesting (plant kingdom) and storing the wood (mineral kingdom).

PHILOSOPHICAL ARGUMENT

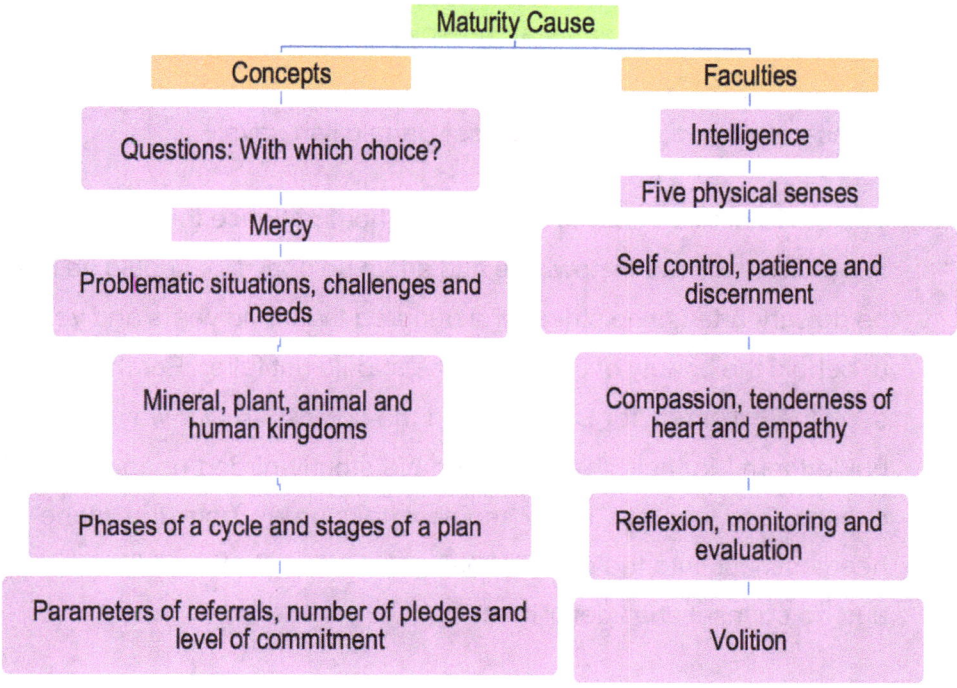

The Cause of Maturity can be systematized and augmented using the following categories in order to organize its most relevant information:

Intelligence

Our capacity to address problems, issues, difficulties and calamities.

Discernment, Self-control and Volition

Search alternatives using our powers of discernment right, and opportunity to choose exercising our free will.

Patience

When we consider that each stage of the process of scientific research implies search, we can start grasping the importance of learning patience since very early in life. In *The Valley of Search* which describes the first phase of human earthly progress, Bahá'u'lláh says:

> The steed of this Valley is patience; without patience the wayfarer on this journey will reach nowhere and attain no goal. Nor should he ever be downhearted; if he strive for a hundred thousand years and yet fail to behold the beauty of the Friend, he should not falter. For those who seek the Ka'bih of 'for Us' rejoice in the tidings: 'In Our ways will We guide them.' In their search, they have stoutly girded up the loins of service, and seek at every moment to journey from the plane of heedlessness into the realm of being. No bond shall hold them back, and no counsel shall deter them.[212]

Human Kingdom

Institutions, government, parents, and everybody in a position of authority

Kingdoms in Nature

Mineral, plant and animal kingdoms and how they are affected by our decisions

---

[212] Bahá'u'lláh, The Seven Valleys and The Four Valleys, p. 6.

Stages and Stations

Stations of a journey and phases or stages of a cycle to explore where the problem is originating and which options seem to be wise to be taken.

Wisdom

Decisions founded in the sources of wisdom and in compassion, tenderness of heart and empathy.

Mechanism of Control

Check the current mechanism of control when going from one stage or phase to the next, or suggesting new ones and also encouraging self-accountability.

Parameters and Indicators of Progress

Such as level of respect for the individual to exercise his (her) free will, level of commitment, number of pledges, and number of recommendations or referrals.

We can perceive that the rational soul, the intelligence and the human spirit are all referring to the same being: one that is free, with duties and rights; that has a free will and can decide to be or not to be responsible, since all the fundamental causes discussed above are subject to laws. All mature individuals are equal before the laws, but in the context of the options taken by each, it is clear that each human being is unique in his (her) potential; in the distinctive characteristics of how he (she) expresses his (her) qualities; in the degree of excellence which he (she) makes manifest his (her) art, profession or trade; in his (her) level of love toward others and of obedience to the norms; and in his (her) noble aspirations and his (her) own weaknesses and limitations.

Evolution

Creation and evolution have been contentious issues between religion and science. In order to address the multiple differences among the contenders I would like to suggest a path that may open a possible common understanding. In the context of the notions relevant to the Cause of Maturity I would like for the reader to reflect on the involuntary differential responses of every living organism during its maturating process.

First, let's try to understand the following quote by 'Abdu'l-Bahá: "Essential pre-existence is an existence which is not preceded by a cause; essential origination is preceded by a cause. Temporal pre-existence has no beginning; temporal origination has both a beginning and an end."[213]

Ian Kluge has some very wise insights:

> Because Aristotle and the Writings do not recognize a hard and fast distinction between physics and metaphysics and / or theology – a fact of enormous significance in our consideration of the unity of science and religion – the Divine is an inevitable part of any discussion of the universe's physical constitution. Not only do both see God as the "Prime Mover" but they also regard God as utterly self-sufficient, meaning, philosophically speaking, as not preceded by a cause or, as the Baha'i Writings say, "Self-Subsisting" and, therefore, independent of all other existing things. According to Aristotle, God is also the First Mover Who is Himself unmoved or unchanged. This is because the Unmoved Mover is pure actuality,

---

[213] 'Abdu'l-Bahá, Some Answered Questions, pp.155 -56. Emphasis added.

that is, has no potentials, and is, therefore, beyond all change because there are no potentials left to actualize. One might also express this by saying that God has no privations, no lacks or deficiencies requiring fulfillment. Moreover, the Divine is one and eternal that is, undivided and beyond time, characteristics which also suggest that God is not in space among other phenomenal beings. God is not limited by the normal attributes of all phenomenal, material beings. God is also alive conscious and thinking.[214]

In *Some Answered Questions,* 'Abdu'l-Bahá tells us that "all creatures emanate from God – that is to say, it is by God that all things are realized"[215], in other words, the potentials of things become real, are real-ized or brought into material existence by God's action. This explains why 'Abdu'l-Bahá considered the movement *from existence into existence* as a *degree* of change, even though Aristotle thought of specifically differentiated *types* of change: for Baha'is, "*generation*", that is, the movement from *non-existence to existence* is simply the change from potentiality to actuality, which is Aristotle's original and fundamental definition of movement. "Alternation" is a change from a something to something else, and in the Baha'i view, the movement from *non-existence*, that is, potential existence, *to existence* is simply the actualization of an already existing potentiality.[216]

Abdu'l-Bahá, The Master, has said:

---

[214] Kluge, Ian. The Aristotelian Substratum of the Baha'i Writings.

[215] Abdu'l-Baha, Some Answered Questions, p. 203.

[216] Kluge, Ian. The Aristotelian Substratum of the Baha'i Writings. Emphasis Added.

All beings, whether large or small, were created perfect and complete from the first, but their perfections appear in them by degrees. The organization of God is one; the evolution of existence is one; the divine system is one. Whether they be small or great beings, all are subject to one law and system. Each seed has in it from the first all the vegetable perfections. For example, in the seed all the vegetable perfections exist from the beginning, but not visibly; afterward little by little they appear. So it is first the shoot which appears from the seed, then the branches, leaves, blossoms and fruits; but from the beginning of its existence all these things are in the seed, potentially, though not apparently.

In the same way, the embryo possesses from the first all perfections, such as the spirit, the mind, the sight, the smell, the taste -- in one word, all the powers -- but they are not visible and become so only by degrees.

Similarly, the terrestrial globe from the beginning was created with all its elements, substances, minerals, atoms and organisms; but these only appeared by degrees: first the mineral, then the plant, afterward the animal, and finally man. But from the first these kinds and species existed, but were undeveloped in the terrestrial globe, and then appeared only gradually. For the supreme organization of God, and the universal natural system, surround all beings, and all are subject to this rule. When you consider this universal system, you see that there is not one of the beings which at its coming into existence has reached the limit of perfection. No, they gradually grow and develop, and then attain the degree of perfection.[217]

---

[217] Abdu'l-Baha, Some Answered Questions, p. 199.

If we accept the above-mentioned quote, that God has created every creature with the potential to express its perfections gradually, then, we could imagine that thru the passing of many natural cycles and the corresponding tests, the opportunity for an organism to express emergent properties begin to be plausible. Because of: variations in the soil or in the air composition; or the micro-climate changed; or enough biodiversity of the surrounding manifested; or its organs were able to establish ideal symbiotic relations with fungus or bacteria. Then believing in "creation" and "evolution" seems to be reasonable if we accept that the "survival of the fittest" is just the consequence of the changes in the conditions of the environment that created the possibility for those features of perfection to gradually appear. Such differential responses happen involuntary. Those changes evolved sequentially in each species over many generations.

In the context of racial discrimination, I have the following questions: Are homogeneous characteristics, within one species, the natural tendency of evolution? Is it a law of evolution that the tendency of the organisms, within one species, is toward diversity of characteristics?

THE QUINTESSENCE OF KNOWLEDGE

The quintessence of knowledge, O my Lord, proclaimeth its powerlessness to know Thee, and perplexity, in its very soul, confesseth its bewilderment in the face of the revelations of Thy sovereign might, and remembrance, in its inmost spirit, acknowledgeth its forgetfulness and effacement before the manifestations of Thy signs and the evidences of Thy praise. What, then, can this poor creature hope to achieve, and to what cord must this wretched soul cling?[218]

---

[218] Bahá'u'lláh, Prayers and Meditations by Bahá'u'lláh, p. 173.

It behoveth him who is a wayfarer in the path of God and a wanderer in His way to detach himself from all who are in the heavens and on the earth. He must renounce all save God, that perchance the portals of mercy may be unlocked before his face and the breezes of providence may waft over him. And when he hath inscribed upon his soul that which We have vouchsafed unto him of the quintessence of inner meaning and explanation, he will fathom all the secrets of these allusions, and God shall bestow upon his heart a divine tranquillity and cause him to be of them that are at peace with themselves. In like manner wilt thou comprehend the meaning of all the ambiguous verses that have been sent down concerning the question thou didst ask of this Servant Who abideth upon the seat of abasement, Who walketh upon the earth as an exile with none to befriend, comfort, aid, or assist Him, Who hath placed His whole trust in God, and Who proclaimeth at all times: 'Verily we are God's, and to Him shall we return'.[219]

*

In conclusion, the cycle of scientific research has also the propositional phase and the notions and the set of human faculties associated in the Maturity Cause are appropriate to be able to answer the question: "With which option?"

Because each cause has its effects, then as we practice the Maturity Cause it will show its effects:

In our understanding of the above-mentioned notions (cycle, phases, stages, kingdoms)

---

[219] Bahá'u'lláh, Gems of Divine Mysteries, p. 25.

In the development of our faculties (patience, discernment, self-control, free will)

In the uses of our powers (intelligence, realizing the existence of two egos, the heart's role)

In the building of relationships (connection between the cycles of nature such as seasons and the plants and animal cycles, the impact of our decisions onto the other kingdoms)

In our understanding of the laws of the cycle related to the problem (For example: illness cycle, domestic violence cycle) and our determination to break the cycle

In the polishing of virtues (detachment, mercy,
    empathy, compassion and tenderness of heart);

In and, our capacity to humbly suggest parameters and indicators to monitor and evaluate the advancement to the solution of any problem.

A similar statement can be expressed about each cause. This approach will be very important in reaping the fruits of a curriculum that teaches the science of the love of God.

X THE FARMERS' SITUATION

In order to help the reader understand the example of the farmers, the author has added emphasis to the key notions of the discourse linked reasonably to each one of the methods.

THE FARMERS' SITUATION IN THE CONTEXT OF THE EXPLORATORY PHASE OF THE CYCLE OF SCIENTIFIC RESEARCH

Let us then start with our example of the farmers with the exploratory phase using the Maturity Cause and then proceed with the four causes taught by the Greek philosophers and finish with the propositional phase employing the Maturity Cause again.

SUMMARY OF THE EXPLORATORY & PROPOSITIONAL METHODS

Question: With which option? when addressing problems, difficulties, issues and needs

Essence: Mercy.

Sources of Wisdom: What history or experience, the diverse disciplines of science and the world religions say about the issue to be addressed

Kingdoms, Cycles and phases

Laws: related to the inherent responsibilities of the assumed decision taking into account the effects in the other Kingdoms and in the Human Kingdom.

Goal: Stewardship in the management of the trust

Powers: Intelligence, patience, self-control, discernment, compassion, empathy, volition

Decisions being made at the individual and collective level: Governmental Institutions, Administration of private enterprises, NGOs, families or individuals. The human decisions in relation to the stages of a plan or phases of a cycle

The spontaneous differential response to the stimuli received by the different kingdoms

The effects of the decisions made by humans in all kingdoms

The developmental stage of the individual, plant or animal

Monitoring and evaluating the strategy and its stages and the cycle and its phases

Relationships: are those that interconnect the sources of wisdom

Relational situations in which the human kingdom is challenge to choose or, in the case of the mineral, plant and animal kingdoms, the appearance of a spontaneous differential response to the stimuli received.

Parameters and indicators of progress: Such as level of respect for the individual to exercise his (her) free will, level of commitment, number of pledges, and number of recommendations or referrals

In the following reflections we can find *choices* made by *governments*, *farmers* and Bahá'u'lláh's instructions to His *Administrative Order*. Let's try to perceive how those *choices* have affected and will affect the *human kingdom, and plant, animal and mineral kingdoms*.

Assume there is a rural population, which has little desire to continue performing agricultural labor. They are unhappy with life in the countryside. Rather, *this population wants to migrate to the city*, with all the consequences that this step implies.

The following is from the Food and Agricultural Organization of the United Nations:

> Small farmers produce much of the developing world's food. Yet they are generally much poorer than the rest of the population in these countries, and are less food secure than even the urban poor. Furthermore, although rapid urbanization is taking place in many developing countries, farming populations in 2030 will not be much smaller than they are today. For the foreseeable future, therefore, *dealing with poverty and hunger in much of the world means confronting the problems that small farmers and their families face in their daily struggle for survival.*

Investment priorities and policies must take into account the immense diversity of opportunities and *problems facing small farmers*. The resources on which they draw, their *choice* of activities, indeed the entire structure of their lives, are linked inseparably to the biological, physical, economic and cultural environment in which they find themselves and over which they only have *limited control*. While every farmer is unique, those who share similar conditions also often share *common problems and priorities* that transcend *administrative or political borders*.[220]

Let us imagine the possibility of *widespread hunger* in a region and the level of *respect towards the government*.

How do you explain to a 4-year-old, who has been playing outside all afternoon, that tonight he is not going to have dinner?

It is important for the farmers to be aware of how acutely will their *descendants* born in the belts of misery of the big cities (Rio de Janeiro, Mexico City, Calcutta, Bogota) be affected in their *IQ*. The author sadly concluded that around *1.2 billion people have been deprived* of a significant portion of their potential *intellectual* capacities because of lack of nourishment of their mothers *during pregnancy* (Duque, 1999) (Joseph) (Martorell).

As an example of *governmental decisions*, the "Food Stamps" program of the United States of America, in spite of the abuses, is worthy of being *evaluated to address this issue*.

---

[220] Dixon, Gulliver and Gibbon, Farming Systems and Poverty: Improving Farmers' Livelihoods In A Changing World. Emphasis added.

In order to *explore* the causes of their situation, small farmers could examine:

[1] *The crop's cycle* and the issues that they had in each *phase* in order to understand which options have to be changed for the next *planting season*.

[2] The harvest *season* of the species farmed.

[3] *The decisions* made based on what happened in the *climatic cycle during the planting, growing and harvesting phases* that may have affected *the mineral kingdom*, for example: lack of enough water to help dissolve the mineral nutrients in enough quantity.

[4] *The decisions* made based on what happened in the *climatic cycle* during the *planting, growing and harvesting phases* that may have affected *the plant kingdom*, for example: an unreasonable growth spurt of *weeds* during the crop's *planting or growing phases*.

[5] *The decisions* made based on what happened in the *climatic cycle* during the *planting, growing and harvesting phases* that may have affected the *animal kingdom* in the soil, for example: due to the intensive use of , herbicides, fungicides, pesticides, chemical fertilizers and heavy machinery during *many cycles*, what has happened to the worms, fungus and microorganisms that used to be in abundance in the soil.

The following is a concrete example if assume that such soils are almost sterile:

> Ecosystem productivity commonly increases asymptotically with plant species diversity, and determining the mechanisms responsible for this well-known pattern is essential to predict *potential changes in ecosystem productivity with ongoing species loss*. Previous studies

attributed the asymptotic diversity–productivity pattern to plant competition and differential resource use (e.g., niche complementarity). Using an analytical model and a series of experiments, we demonstrate theoretically and empirically that host-specific soil microbes can be major determinants of the diversity–productivity relationship in grasslands. In the presence of soil microbes, *plant disease* decreased with increasing diversity, and productivity increased nearly 500%, primarily because of the *strong effect of density-dependent disease on productivity at low diversity*. Correspondingly, *disease was higher in plants* grown in conspecific-trained soils than heterospecific-trained soils (demonstrating host-specificity), and productivity increased and host-specific disease decreased with increasing community diversity, suggesting that *disease was the primary cause of reduced productivity in species poor treatments. In sterilized, microbe-free soils, the increase in productivity with increasing plant species number was markedly lower than the increase measured in the presence of soil microbes*, suggesting that niche complementarity was a weaker determinant of the diversity– productivity relationship.

Our results demonstrate that soil microbes play an integral role as determinants of the diversity–productivity relationship.[221]

[6] *The decisions* made based on what happened in the climatic cycle during the *planting, growing and harvesting phases* that may have affected the *animal kingdom*, for example: The *pest management* practices and the effect in beneficial insects that serve as predators of *damaging insects*.

---

[221] Schnitzer, Stefan A. et al. Emphasis added.

[7] The historical *decisions* made by those in *authority* in relation to the rural areas and agriculture. For a sample of how governmental power has been exercised and imposed, see Brocket's Land Power and Poverty. Agrarian Transformation and Political Conflict in Central America, which also covers the positive experience of South Korea in this regard.

In order to come up with a strategic solution and increase their own understanding and be *empowered* to presented it to those in *authority*, they could use a macro-economic forecasting software, to determine the implications and *the phases and stages that they have to monitor to follow up the decisions required*, such as: decreasing the use of *pesticides* and opt for an integrated *pest management strategy*, rotating polycultures instead of permanent monocultures, decreasing fertilizers applications and *opting* for composting and natural fertilizers, and to minimize water usage, prevent *soil erosion* towards a sustainable agriculture and *evaluate* if the farmers are to rely more or less on income sources outside the farm and how much of the aggregated value goes back to the farmers pocket.

[8] In spite of the *difficult situation I choose* being inspired by Religion as a *source of wisdom*. For over two thousand years Christians have been praying:

> "'Our Father in heaven, hallowed be your name, Your *Kingdom* come, Your *Will* be done, on earth as it is in heaven. ..."

In the Bible:

> And it shall come to pass in the last days, that the mountain of the *LORD's house* shall be established in the top of the mountains, and shall be exalted above the hills; and all nations shall flow unto it.

> And many people shall go and say, Come ye, and let us go up to the mountain of the LORD, *to the house of the God* of Jacob; and He will teach us of His ways, and we will walk in His paths: for out of Zion shall go forth the law, and the word of the LORD from Jerusalem.
>
> And He shall judge among the nations, and shall rebuke many people: and *they shall beat their swords into plowshares, and their spears into pruning hooks: nation shall not lift up sword against nation, neither shall they learn war any more.*[222]

In a statement, Shoghi Effendi mentions that the aim of Bahá'u'lláh is to reconcile *conflicting*[223] *creeds*:

> His mission is to proclaim that *the ages of the infancy and of the childhood* of the human race are past, that *the convulsions* associated with the present *stage of its adolescence* are slowly and *painfully* preparing it to attain the *stage of manhood*, and are heralding the approach of that *Age of Ages when swords will be beaten into plowshares, when the Kingdom* promised by Jesus Christ will have been established, and the peace of the planet definitely and permanently ensured.[224]

In the Bahá'í Writings:

> Whilst in the Prison of 'Akká, We revealed in the Crimson Book that which is conducive to the advancement of mankind and to the

---

[222] Isaiah, 2:2 –2:4. Emphasis Added.

[223] Is a problem

[224] Summary Statement - 1947, Special UN Committee on Palestine. Emphasis Added.

reconstruction of the world. The utterances set forth therein by the Pen of the Lord of creation include the following which constitute the fundamental principles for the *administration* of the affairs of men:

First: It is incumbent upon the *ministers of the House of Justice* to promote the Lesser Peace so that the people of the earth may be relieved from the burden of exorbitant expenditures. This matter is imperative and absolutely essential, inasmuch as *hostilities and conflict* lie at the root of *affliction and calamity*.

Second: Languages must be reduced to one common language to be taught in all the schools of the world.

Third: It behoveth man to adhere tenaciously unto that which will promote fellowship, kindliness and unity.

Fourth: Everyone, whether man or woman, should hand over to a trusted person a portion of what he or she earneth through trade, agriculture or other occupation, for the training and education of children, to be spent for this purpose with the knowledge of the *Trustees of the House of Justice*.

Fifth: Special regard must be paid to agriculture. Although it hath been mentioned in the fifth place, unquestionably it precedeth the others. Agriculture is highly developed in foreign lands, however in Persia it hath so far been grievously neglected. It is hoped that His *Majesty the Shah* -- may God assist him by His grace -- will turn his attention to this vital and important matter.[225]

---

[225] Bahá'u'lláh, Tablets of Bahá'u'lláh, pp. 89 – 90. Emphasis Added.

Arnold Toynbee in his "Study of History" thoroughly explains the relationship between the Law and the exercise of the *free will* in History; and demonstrates in an irrefutable way the submission of the human affairs to the laws of nature. When explaining the *collapse of the Syrian, Indic and Hellenic civilizations* he mentions *lack of food security* and the deterioration of the population's health *when agriculture was disrupted* in some critical regions. He also said in clear terms that corruption of religion is the cause of the falling of civilizations. Are these not the same reasons why the *actual civilization is falling down*?

THE FARMERS' SITUATION IN THE CONTEXT OF THE FORMATIVE PHASE OF THE CYCLE OF SCIENTIFIC RESEARCH

Let us then continue with our example of the farmers using the Formative Method.

SUMMARY OF THE FORMATIVE METHOD OF SCIENCE

Question: How is the arrangement?

Essence: love and soul as its power.

The general properties of matter: form, size, mass, temperature, movement and position in time and space

Hormones, feelings and emotions

Laws related to unity, harmony, justice and the law of gravity.

Goal: Aspirations and desires to reach the expected vision or design

Powers: imagination and senses of justice and religion

The practical application of the principles related to the general properties of matter: form, size, mass, temperature, movement and position in time and space in relation to the desire organization, arrangement and design

Religious and Scientific principles related to the general properties of matter

Parameters and indicators of unity, beauty, symmetry, harmony, reciprocity and justice in reaching the desire vision or design

Some historical facts about farmers and merchants:

> The aristocratic Athenian State was based upon land-ownership, slavery, and the *entire* freedom of the land-owning class from *all* but family and State duties, from *all* need of engaging in productive industry. So long as the chief wealth of the State consisted in land and its produce, so long as the population was *divided into two classes, the rich and the poor*, and so long as the former had *little* difficulty in keeping *all power* in its own hands. But no sooner did the *growth of commerce* throw wealth into the hands of a class that owned *no land*, and was not above engaging in industry, than this class began to claim a *share* in political power.
>
> There were *now two wealthy classes, standing opposed to each other*, a proud, conservative one, with "*old wealth and worth*," and a vain, radical one, with *new wealth and wants*, both bidding for the favor of the class that had *little wealth, little worth, and many wants*, and thus making it feel its importance. Such is the origin of Athenian democracy.
>
> It is the child of trade and productive industry. It owed its final consecration to the Persian Wars, and especially to the battle of Salamis, in which Athens was saved by her fleet, manned chiefly by marines from *the lower classes, the upper classes*, as we have seen, being trained only for land-service. Thus, the battle of Salamis was not only a victory of Greece over Persia, but *of foreign trade over home agriculture*, of democracy over aristocracy. [226]

---

[226] Aristotle and Ancient Educational Ideals by Thomas Davidson. Chapter V: Education as influenced by time, place, and circumstances. Emphasis added.

Let us continue with our example. To understand how a rural population became so disillusioned with the viability of its farming businesses and the reasons it migrated to the misery belts around big cities, we should analyze:

[1] The principles of preservation of natural resources and the *soil topography in the region and in the farm*.

[2] The principles of agriculture and crop layout, especially in relation to environmental sanitation and the *disposition of sick plants and farm waste*.

[3] The correlation between the *various kinds of discrimination* (dark skin, women, poorly educated, poorly dressed and of course being part of the farmers' social class) and its *cumulative effect* on *how much* society values what is produced.

The International Labor Organization (ILO), which is a United Nations agency, says:

> *About 40 percent of the world's three-billion strong labor force, some 1.2 billion workers*, are employed in agriculture as self-employed farmers, unpaid family workers and hired workers. The ILO puts the number of 'waged' or hired workers at *450 million, 38 percent of all persons employed in agriculture* and *equivalent to the entire labor force of the high-income countries*. The ILO includes as waged workers those who receive in-kind payments, and notes that some workers have *several statuses*, such as a person who is a farmer at *some times of the year* and hired worker at others.
>
> ... *Many of the world's hired farm workers* are employed on plantations, *some owned by multinationals*, that provide housing and wages according to the terms of collective bargaining agreements.

Plantation workers in 'breakfast commodities' such as banana, coffee and sugar are *sometimes among their countries'* working elite. However, *some multinational producers of breakfast commodities* have replaced *year-round workers* who had housing, schools and other amenities on the plantation with women hired seasonally and offered few or no benefits.[227]

Rural women constitute *50 percent of the agricultural labor force in Africa*; they are responsible for *80 percent of the food production and 50 percent of the agricultural output.* Women reinvest almost *90 percent of their income* in their children and household. Since women are the keys to improving household food security and nutritional wellbeing, *increasing women's access to financial resources* ultimately leads to *increased investments* in human capital.[228]

Proponents of the current model of economic production will have to answer whether they are being coherent to the *principles of economics when they consider that at least 1.2 billion farm workers* are being affected by the problem of *discrimination in one or more ways*.

In a document titled Gender and Value Chain Development we find:

A gendered impact assessment of organic certified pineapple and coffee producers in Uganda (Bolwig and Odeke, 2007) also touches open changes in women's *workload* as a consequence of certification. The study, based on a household survey (with control

---

[227] ILO. Global Farm Worker Issues. Emphasis added.

[228] Kwelagobe. Commentary - Investing in Rural Women: Closing the Gender Gap in African Agriculture. Emphasis added.

group) and focus group interviews, discusses how organic conversion affects men and women differently in respect of changes in the costs and benefits of farming. The study finds that organic conversion has significantly *increased women's labour effort* in coffee production, while the effect on male labour has been weaker. While men enjoy almost exclusive control of income from organic farming, *it is women who carry out most of the additional farming and processing work* needed to meet organic certification and stricter quality and farm management requirements of the organic exporter. According to the authors, it is very likely that *women's increased effort in coffee farming in recent years* has occurred at the expense of their own income-generating activities. Hence, while men over the *last five years* have enjoyed an *increase in the income they control* (from coffee), *women appear to have experienced the opposite*. The skewed gender distribution of the costs and benefits associated with certification was much more pronounced among coffee farmers. According to the authors, this seemed to be the result of *differences in gender relations*, in land availability and *farm size*, and in market conditions: 1) *gender relations seemed generally more equal among pineapple farmers* thus giving *women better access to pineapple incomes* and *men less command* over women's labour for the purpose of pineapple growing; 2) pineapple farmers earned *very high incomes (in local comparative terms)*, due to *larger farm size, high yields*, and favourable market conditions, and this allowed them *to hire more labour* thereby relaxing the demand on women's labour.[229]

---

[229] Ministry of Foreign Affairs of Denmark. Gender and Value Chain Development. Emphasis added.

In the *same* way that much of the early Green Revolution literature (Feder and O'Mara 1981) focused on limited *small farmer* uptake of improved seeds, fertilizer, irrigation, and other components of "modern" production systems, a *large share* of the emerging literature on modern value chains has been concerned with *smallholder participation* in AVCs (Agricultural Value Chains)[230] and with whether these *same value chains* might be leaving *many poorer farmers behind.*[231]

This is perhaps unsurprising given that, historically, market sales of food have been *heavily concentrated* in the hands of a *small number of producers*, even in *regions and countries in which market participation is broad-based*. Although most of the evidence comes from staple grains markets, a *relatively small group* (i.e., *less than 10 percent*) of relatively well-capitalized farmers *located in more favorable agro-ecological zones account for a significant majority of market sales throughout the world* (Barrett 2008). This suggests that gains from agrifood value chain transformation accruing to net sellers in the form of *higher profits will likely concentrate in the hands of a relatively modest share of the farm population* in the developing world, although there is presently *scant* hard evidence on this important point.[232]

---

[230] AVC stands for: Supermarkets, specialized wholesalers, and processors and agro-exporters' agricultural value chains have begun to transform the marketing channels into which smallholder farmers sell produce in low-income economies.

[231] Christopher B. Barrett, Smallholder Participation in Agricultural Value Chains: Comparative Evidence from Three Continents. p. 32. Emphasis added.

[232] idem. p. 32. Emphasis added.

Gender and generic value chain interventions:

Generic value chain interventions can have *positive effects for participating women*. But this is more likely to happen when they take into consideration gendered constraints that apply to *upgrading value chain participation and distribution of value* (both along the value chain and within households). For example, value chain interventions emphasizing product and process upgrading as well as forging/strengthening *horizontal and vertical linkages*, can be strategically applied in *parts of the chain* where women play important roles. But in order to secure *positive impacts* and avoid unintended *negative consequences*, a gendered value chain analysis is needed as part of *project design* and implementation. Assuming that women will automatically *gain* from generic value chain interventions can have unintended *negative consequences*. Furthermore, these interventions are not sufficient in themselves to *secure value chain participation* and meaningful welfare outcomes for women when working in 'gender conservative' areas. Finally, the gender impact of generic value chain interventions is likely to be *mainly limited to improving the terms of inclusion* of existing value chain participants, rather than promoting the *participation of more women* in the chain.

Generic value chain interventions can (in specific circumstances) have *positive effects* for participating women; but in order to secure *positive impacts* and avoid unintended *negative consequences*, a gendered value chain analysis is required as part of *project design* and implementation.[233]

---

[233] Ministry of Foreign Affairs of Denmark. Gender and Value Chain Development. Emphasis added.

The reality of women farmers *not being paid what they deserve* is not just common among the *masses of the poorly educated*. Academia is also responsible of this situation. An example from personal experience comes to mind. When speaking to a PhD in economics, it was stated that he firmly believed that work done in the household by my wife cannot be *counted* as such, because there is not an *economic* transaction. I tried several ways to convince him that housework brings wellbeing, but he kept saying that it could not be perceived as work from an *economic perspective*. After several attempts, I told him that from the physics perspective work is movement and energy consumption, which finally convinced him. It is my feeling that *Fragmentation* of knowledge into disciplines has been the cause of *many divisions* in society.

[4] The principles of the rural economy at the regional and local level and the organization of the market.

Let's see:

"The yield of the coffee tree *peaks after 5 to 7 years*. The fruits are left unpicked until they reach the ideal stage of ripeness, *usually after about seven months*."[234]

> "Coffee has *low supply elasticity* just as the case with cocoa due to the perennial nature of the crops and the *demand is very inelastic*. A situation of *supply shortage* results in *high coffee prices* without a *significant reduction in consumption*; and *when prices are high* it takes *time* for production to adjust. This is exacerbated by the lag

---

[234] Traoré Cocoa and Coffee Value Chains in West and Central Africa: Constraints and Options for Revenue-Raising Diversification. p. 50. Emphasis added.

between plantation and harvest, *which varies between 18 and 24 months* for coffee."[235]

When 'Abdu'l-Bahá was asked about the solution to the *economic* problem he is reported to have said: "The solution of this problem is one of the fundamental principles of His Holiness Bahá'u'lláh. But it must be solved with *justice* and not with force. If this problem is not solved *lovingly* it will result in war."[236]

In order to propose a solution, a macro-economic forecasting software and other means could be used to examine the implications for farmers and the world as a whole of a "*uniform* and universal system of currency" where "the economic resources of the world will be *organized*, its sources of raw materials will be tapped and *fully* utilized, its markets will be *coordinated* and developed, and the *distribution* of its products will be *equitably* regulated."[237]

[5] Now, compare the complexity of the *number of variables* that a farmer deals with, to those handled by the agroindustry that processes and packs the food and to the *limited number of variables* that merchants deal with.

There are over one million species of insects, *around 298,000* species of plants, and around *611,000* species of fungus" (Mora, Camilo et al. How Many Species on Earth and in the Ocean?), and there are *many* species of bacteria and viruses. Those who labor on a farm are supposed to be trained

---

[235] idem. p. 55. Emphasis added.

[236] Compilations, Bahá'í Scriptures. p. 340. Emphasis added.

[237] Shoghi Effendi in the Introduction of a book by Bahá'u'lláh, The Proclamation of Bahá'u'lláh, pp. xi-xii. Emphasis added.

to recognize the different kinds of pests that may affect the crops and to report them *immediately*.

[6] Also, they should try to correct agriculture commodities distortions such as when there is a small number of buyers (oligopoly).

In a document of the Food and Agriculture Organization of the United Nations, we find:

> West and Central Africa produces about 70 percent of world cocoa. About 90-95 percent of all cocoa is produced by smallholders with farm sizes of two to five hectares (Ha).
>
> Big buyers can pick and choose, playing one producer country against the other. In Cote d'Ivoire just three years after liberalization there were forty registered exporters, but ten control over 90 percent of the market. Legislation prevents market shares of these exporters from increasing. Concentrated exporters can potentially exercise market power both on farmers and traders in the producing countries and on manufacturers in the consuming countries. ...
>
> Three Transnational Corporations now dominate the processing and supply of the intermediate cocoa product (cocoa butter and powder, and 'industrial' chocolate), accounting for over 35 percent of total worldwide cocoa grinding capacity (Talbot, 2002).
>
> The continuing strong performance of Nestlé and other giants on the processed beverage world is in outstanding contrast with the ever-increasing impoverishment of ordinary coffee farmers at a time of low green coffee prices. ...
>
> Multinationals capture most of the value-added linked with the production of cocoa and coffee. To secure their market share and increase their profit margins, they have made huge investments in

branding and advertising, which shelters them from price competition. While coffee prices almost halved between 1999 and 2001, the average retail prices in the US (the largest consumer in volume) decreased by less than 4 percent (Ponte, 2001). This suggests that not only gross margins have increased for roasters, but also profits."[238]

"The relatively greater success for coffee value chain can be attributed to several factors including the fact that consumers buy coffee beans directly, whereas cocoa beans are used as ingredients in recipes and *never purchased* directly by consumers. A second difference is that *there is more Transnational Corporations* involvement in cocoa processing *located in the producing countries than is the case of coffee.*"[239]

With the same forecasting software, examine the consequences of paying women and other groups discriminated against what they fairly deserve and the results of fair trade

---

[238] Traoré Cocoa and Coffee Value Chains in West and Central Africa: Constraints and Options for Revenue-Raising Diversification. p. 52. Emphasis added.

[239] idem. p. 50. Emphasis added.

THE FARMERS' SITUATION IN THE CONTEXT OF THE EXPLICATIVE PHASE OF THE CYCLE OF SCIENTIFIC RESEARCH

Let us then continue with our example of the farmers using the using the Explicative Method.

SUMMARY OF THE EXPLICATIVE METHOD OF SCIENCE

Questions: Why? For what?

Essence: the power of law itself, i.e., that which is a condition to a certain aspect of life or nature, and explains its purpose, its mission

Social order, laws and norms; Natural laws; Religious laws and ordinances. Pacts, agreements and covenants. Individual and collective commitment to ideals

Goal: the end, the mission, the purpose

Powers: Memory and senses of responsibility, fear and shame

Advice from those with experience. The **consequences** of obeying and disobeying.

Situations of risk and danger.

Relationships of gratitude, loyalty, reciprocity, mutuality, of fear and protection; and also guilt, repentance, punishment and reward.

Parameters and indicators of protection, equality, prevention and security in the fulfillment of the mission

Let us illustrate the final cause continuing with the example of the farmers. The deepest causes for their attitudes towards peasant life can be found if they examine:

[1] The *sense of fear of going bankrupt*, because of the *vulnerability* of the crop once it has been harvested without having the means to protect it from decomposing. For example:

> "In Akuapem South district, pineapples are a high-value, nontraditional crop grown primarily for export as whole fruits. As described in Conley and Udry (2010), the opening of European pineapple markets to Akuapem farmers in the mid-1990s had a transformative effect on local agriculture. But, as Fold and Gough (2008) describe, unanticipated changes in the European market around 2004 *caused major disruptions* for Ghanaian pineapple growers and *fundamentally altered the terms of their contracts. Verbal agreements were not honored*, and in some cases firms which had begun the process of harvesting pineapples from smallholder farms *neglected to return to pick up the fruit, leaving the farmers with unsellable produce and without payment*. Both farmers and exporting firms *lost their businesses* as a result of the demand shock, leading to a period of intense rationalization in the industry. Farmers interviewed in 2009 *expressed regret for accepting verbal contracts* with the buying firms, and reported that they would *no longer sell without a written and legally binding agreement* (Harou and Walker 2010)."[240]

Despite having to pay a cooperative membership fee in Ghana, most Ghanaian pineapple farmers join because these groups have greater bargaining power, the *ability to demand written contracts and the financial might to take legal action in response to breach of contract.*

---

[240] Christopher B. Barrett, *Smallholder Participation in Agricultural Value Chains: Comparative Evidence from Three Continents.* p. 17. Emphasis added.

Cooperatives are also a vehicle for accessing resources and skills training. In Ghana, 27 percent of cooperative members mentioned the increased likelihood of receiving help from the government or from an NGO as their main reason for joining a cooperative.[241]

"From our observations across the case study countries, the problem of *holdup* by firms appears to increase in the number of smallholders with whom the firm *contracts*. As firms face a larger pool of prospective suppliers, especially when the *contract product is perishable*, firms appear more likely to speciously reject commodities as not meeting *agreed quality standards*, or simply not show up to *purchase contracted commodities*. In Ghana, firms and their middlemen commonly come to harvest the crop. If they d*o not show to harvest,* collect and pay for the crop, the smallholder's only outside option is sale on the local market at a much lower price, roughly half, or *outright loss due to spoilage caused by waiting on the contracting firm*. Similar problems were observed in India and Nicaragua in horticultural products."[242]

[2] Their understanding of, and attitude toward, natural laws, which teach us, for example, that only plants are direct collectors of the sun's energy and all other organisms are indirect consumers.

[3] The social laws concerning protection of peasants and agriculture from multiple risk and other factors, such as Water Rights for farmers.

For example:

---

[241] idem. p. 27. Emphasis added.

[242] idem. p. 29. Emphasis added.

The European Economic Community was designed to create a common market among its members through the elimination of *barriers* and the establishment of a *common external trade policy*, The *treaty* also provides for a *common trade agricultural policy*, which was established in 1962 to *protect EEC farmers from agricultural imports*. The first reduction in *EEC tariffs* was implemented in January 1959, and by July 1968 *all internal tariffs were removed*. Between 1958 and 1968 trade, among the EEC's members, quadruple in value.[243]

The International Labor Organization says:

In several countries the *fatal accident rate in agriculture is double the average for all other industries.* ILO estimates that workers *suffer 250 million accidents* every year and out of a total of *335,000 fatal workplace accidents worldwide, there are some 170,000 deaths among agricultural workers.*[244]

[4] The banking system and the commercial laws and regulations affecting loans to farmers.

[5] The recommendations and suggestions of those with experience.

[6] Religious teachings, which in this case say:

First and foremost is the principle that to all the members of the body politic shall be given the greatest achievements of the world of

---

[243] Encyclopedia Britannica. Emphasis Added.

[244] ILO Statement to the 56th Commission on the Status of Women. Adoption of international labour standards key to supporting rural women. Emphasis added.

> humanity. Each shall have the utmost welfare and well-being. To solve this problem we *must* begin with the farmer; there will we lay a foundation for system and *order* because the peasant class and the agricultural class exceed other classes in the importance of their service. In every village there *must* be established a general *storehouse* which will have a number of revenues.[245]

This *storehouse* is supposed to play an important role in *providing food to the poor* and in *protecting the farmer* if he (she) *loses the harvest*—so long as it is not through his (her) own negligence.

[7] To my understanding, *food security* and the *protection* of our health and the environment and its biodiversity should be considered the foundation of *order*.

Let us briefly take into account the *risks* involved in being a farmer; *risks* associated with the environment, pest control, insecticides and price fluctuations. Compare those *risks* to those assumed by the agribusiness that transforms the cereal into oats or flour and the merchant who sells the product; and the profits received by each of them. Have you heard about the coffee farmers and their income in third world countries? Tell me now why most of the farmers of the world receive very low income, if one of the basic principles of economics states that *risk* and profit are directly related: more *risk* exposure should lead to higher profit, and vice versa?

> It is believed that the link of the supply chain that is the closest to the farm-gate may be the least competitive one; as *cash trapped farmers* in remote areas *lack good market information and encounter*

---

[245] 'Abdu'l-Bahá, Foundations of World Unity, p. 39. Emphasis Added.

relatively few buying agents[246] (Wilcox and Abbott, 2006). Despite the apparent importance of government support, few producer countries have policies that provide small farmers with a level of playing field. Scale economies in processing, marketing, and distribution as well as market power may lie behind the larger observed margins. A lack of competition along the cocoa supply chain means that farmers capture as little as *0.5 percent of the retail price of cocoa*. Small farmers, contrary to plantations, *are rarely able to by-pass intermediaries* [247] as they do not have basic processing or transportation facilities. In addition, small producers *do not have good access to international price information* [248], which enables local traders to take bigger margins. Finally, farmers *cannot chose the timing of their sale* as they lack access to credit or warehousing facilities, and often have to sell their harvest in advance to cover immediate expenses. High marketing costs such as in-country transportation reduces the share captured by farmers. Farmers living in producing regions far away from any export point, for instance in big and *landlocked countries*, are bound to receive a lower price than farmers close to a sea port." [249]

"Because of the way the international coffee supply chain works, the link between producers and consumers is lost (Oxfam 2002a). Coffee is traded down a complex line of intermediaries, ranging from local traders, exporters, international traders, roasters and retailers, who

---

[246] These are dangerous exposures

[247] Meaning more vulnerability

[248] Idem.

[249] Traoré Cocoa and Coffee Value Chains in West and Central Africa: Constraints and Options for Revenue-Raising Diversification. p. 27-28. Emphasis added.

each capture a percentage of the retail value of coffee. Less than 30 percent of the revenues generated by world coffee sales remains in the coffee producing countries and smallholders usually capture *less than 10 percent of the retail price.*"[250]

The situation at the local market is similar:

It was revealed at the Aruligo meeting that all of the farmers present sold their pineapple as whole fruit at the market.

When asked about processing the fruit into slices none of the farmers present indicated that they do this.

During market investigations it was revealed that there were middlemen who were buying fruit at reduced prices early in the marketing day and then taking these fruits to smaller markets or their own settlements and processing them into slices. The small pineapple were being purchased by non-market vendors for as low as $1 per fruit and then taken to their settlements or commercial areas, cut in half and sold for $2 per slice. This same size was available at the market later in the day as a whole fruit for around $5. Larger pineapples were being sold as whole fruit for around $10-15 each.

The sale of pineapple slices at the Honiara central market was only observed at one stall. This stall was selling whole pineapple fruit from Malaita and pineapple slices that were kept in a sealed plastic container and sold for $2 per slice; it appeared that the same size fruit used that was cut was available whole for around $5. For this case it

---

[250] idem. p. 51. Emphasis added.

is not quite clear why the farmer chose to peel and slice his product only to sell it for less than he was selling his whole fruit for. More investigation into this activity at the Honiara Central Market would be worthwhile to determine the profit margins and if it is a good diversification activity for Aruligo market vendors.

It would appear that this type of practice is most common during the peak pineapple season when the market is full of pineapple and the price drops significantly, however this would have to be validated by further market data collection. One Aruligo farmer informed us that even when the market is very full of pineapple he has never seen anyone through fruit away at the end of the day. Vendors who live far from Honiara will often sleep at the market for as many days as necessary to sell their produce.251

If the population and the institutions *agree that their mission* is to provide a sustainable food source, then they must make the utmost effort to *abide by the norms and laws agreed* and adjust the existing *legislation* to reflect these new *pacts and agreements*.

In order to come up with a proposed solution farmers and others interested could use a macro-economic forecasting software to determine the *consequences* of what will happen if the farmer makes more profit than merchants and those who pack or transform the produce. The marginal profitability for the farmers must be in proportion to *the risk* of each of the actors involved and also the measures taken to *protect our health and the ecosystem*. I do believe that in many cases artisans and those who transform raw materials into final products have *greater risks* and more labor involved

---

[251] Kyle Stice. Aruligo Pineapple Value Chain – Mapping Report. pp. 6-7. Emphasis added.

than those of the merchant. It is simply the application of the principles of economics in relation to the amount of labor and the amount of *risk involved in farming a commodity*, its transformation, and its marketing; and how much of the aggregated value of the final product stays in the hands of the farmer. My *advice* is to gradually apply these changes.

Also as a matter of analysis with the macro-economic forecasting software the role of the *storehouse* should include how it will handle situations of scarcity and abundance and at what price will it buy the food to provide for the poor. The storehouse should buy from the farmers at the price fixed by *the laws of supply and demand*.

'Abdu'l-Bahá said in reference to the revenues of the *storehouse* the following:

> As to the first, the *tenths or tithes*: we will consider a farmer, one of the peasants. We will look into his income. We will find out, for instance, what is his annual revenue and also what are his expenditures. Now, if his income be equal to his expenditures, from such a farmer nothing whatever will be taken. That is, he will not be subjected *to taxation* of any sort, needing as he does all his income. Another farmer may have expenses running up to one thousand dollars we will say, and his income is two thousand dollars. From such an one a *tenth* will be required, because he has a surplus. But if his income be ten thousand dollars and his expenses one thousand dollars or his income twenty thousand dollars, he will have to pay as *taxes*, one-fourth. If his income be one hundred thousand dollars and his expenses five thousand, one-third will he have to pay because he has still a surplus since his expenses are five thousand and his income one hundred thousand. If he pays, say, thirty-five thousand dollars, in addition to the expenditure of five thousand he still has sixty thousand left. But if his expenses be ten thousand and his income two hundred thousand then he *must* give an even half because ninety

thousand will be in that case the sum remaining. Such a scale as this will determine allotment of *taxes*. All the income from such revenues will go to this general *storehouse*.

Then there must be considered such *emergencies* as follows: a certain farmer whose expenses run up to ten thousand dollars and whose income is only five thousand, he will receive necessary expenses from the *storehouse*. Five thousand dollars will be allotted to him so he will not be in need.[252]

Prior to continuing with this search and a proposal they should invite all farmers to a community gathering to established a collective pact based on the warning of the Master, who is reported to have said: "The solution of this problem is one of the fundamental principles of His Holiness Bahá'u'lláh. But it must be solved with justice and not with force. If this problem is not solved lovingly it will result in *war*."[253]

To *avoid war*, my recommendations are: Always avoid debates and encourage consultation processes. Invite *all the parties* (representatives of the government, industry and business sectors, farmers, artisans, merchants and academia) so they grasp with their own understanding the outcome when applying the macro-economic forecasting software. Have at least the same number of women and men participating in the consultation process[254], but a majority of women will help to *ensure a peaceful end result*.

---

[252] 'Abdu'l-Bahá, Foundations of World Unity, p. 40. Emphasis added.

[253] Compilations, Bahá'í Scriptures, p. 340. Emphasis added.

[254] The Universal House of Justice, The Promise of World Peace, p. 6. . Emphasis added.

THE FARMERS' SITUATION IN THE CONTEXT OF QUALITATIVE PHASE OF THE CYCLE OF SCIENTIFIC RESEARCH

Let us then continue with our example of the farmers using the Qualitative Method.

SUMMARY OF THE QUALITATIVE METHOD OF SCIENCE
Essence: The Spirit,
Elements, substances and raw materials.
Specific properties of matter, peculiarities of things and characteristics.
Moral values, also called spiritual values.
Knowledge, beliefs and social values
Spiritual laws, the standards that regulate quality and laws of possession
Goals and objectives
Powers: conscience, knowledge, common faculty and senses of spirituality and appreciation
Specific lines of action or policies to consolidate ethics, knowledge, quality and efficacy
Relationships of belonging, possession, distinction and characterization
Parameters and indicators of efficacy and quality in attaining the goals and objectives

Continuing with our example and with the *conviction* that it is possible to find the profound reasons why this rural population is *adamant* believing that it is not viable to keep trying to survive laboring the land, we should consider the following questions: "What did we have?" "What do we have?" "What do we want to have?" and, "What do we need to achieve it?"

To answer these questions, we have to examine:

[1] The cultural and spiritual values about the worth of the farmers' work.

[2] Their knowledge of the specific properties of the species being produced and of those in the environment.

[3] Their knowledge about the specific properties of soil, natural resources, elements, substances, raw materials, equipment, nutrients, products, services and wastes pertaining to agriculture and cattle-raising.

[4] The cultural and spiritual values, convictions and knowledge about agriculture transmitted by the educational establishment.

[5] The policies and budgetary dispensations of the Ministry of Agriculture[255].

[6] The availability of affordable financing and credit for farmers.

I suggest to the reader to appreciate in the following appraisal the way politicians value farmers, especially reflecting on the needs of small ones in a country and at the world at large:

> Latha Reddy Musukula was making tea on a recent morning when she spotted the *money lenders* walking down the dirt path toward *her house*. They came in a phalanx of 15 men, by her estimate. She *knew their faces*, because they had walked down the path before.
>
> After each visit, her husband, a farmer named Veera Reddy, sank deeper into silence, frozen by some terror he would not explain. Three

---

[255] It is obvious that there will be government policies related to the five causes, for example: policies of plant sanitation and environmental protection; pricing policies; commercialization macro-economic policies; transference of technology and educational policies, and regulatory policies to allow fish reproduction cycle.

times he cut his wrists. He tied a noose to a tree, relenting when the family surrounded him, weeping. In the end he waited until Ms. Musukula stepped out, and then he hanged himself from a pipe supporting their roof, leaving a *careful list of each debt he owed* to each money lender. She learned the full sum then: 400,000 rupees, or $6,430.

…. . India's small farmers, once the country's *economic backbone and most reliable vote bank,* are increasingly being left behind. With global competition and rising costs cutting into their lean profits, their ranks are dwindling, as is their contribution to the gross domestic product. If *rural voters once made their plight into front-page news* around election time, this year the large parties are jockeying for *the votes of the urban middle class*, and the farmers' voices are all *but silent.*

Even death is a stopgap solution, when farmers like Mr. Reddy take their *own lives, their debts* pass from husband to widow, from father to children. Ms. Musukula is now trying to *scrape a living* from the four acres that defeated her husband. Around her she sees a country transformed by economic growth, full of opportunities to break out of *poverty*, if only her son or daughter could grasp one.

But the trap that closed on her husband is tightening around her. Like nearly every one of her neighbors, she is locked into a bond with village money lenders — an intimate bond, and sometimes a menacing one. No sooner did they cut *her husband's body* down than one of them was in her house, threatening to block the cremation unless she paid.

Her appeals to officials for help have been met with indifference. Lately, her fear has been getting the better of her.

'Sure, they will pay, otherwise it would be as if someone has broken into our house and stolen *our money*,' said Sudhakar Ravula, a slight man who lives in a village about two miles away. He introduces himself as a fisherman, but, under questioning, fishes out a pair of gold-rimmed reading glasses and unfolds a *promissory note* signed by Veera Reddy.

Four years ago, he said, he used *borrowed money to lend Mr. Reddy $800*, at an annual interest rate of 24 percent. Reminded of Mr. Reddy's suicide, Mr. Ravula looked impatient. 'I always feel sad for the man,' he said, 'but committing suicide is not the right way to go about it'.[256]

Fifty years ago, the majority of the world's population *lived in rural areas*. Once the majority of the population was force to migrate from the rural areas by different factors, such as political conflicts, violence, drug trafficking, natural disasters, high interest rates and development models, politicians diminished the funding to schools, institutions and infrastructure. They shifted the *priorities to urban areas and the farmers' votes became irrelevant* and their sufferings *invisible*.

My hypothesis is that the majority of politicians and economists seem to share the same theory of economic value in issues such as:

> [a] In USA "In 2007, 1.73 billion tons of *topsoil was lost to erosion*, equal to about 200,000 tons each hour."[257]

---

[256] Barry, Ellen., After Farmers Commit Suicide, Debts Fall on Families in India. Emphasis added.

[257] Michigan University, Center for Sustainable Systems. Emphasis added.

[b] "Despite a tenfold increase in insecticide use between 1945 and 1989, *crop losses due to insect damage nearly doubled.* In 2007, the U.S. agriculture sector used 877 million pounds of pesticides."[258]

[c] "*Nutrient runoff* in the agricultural upper regions of the Mississippi River creates a *hypoxic "dead zone"* in the Gulf of Mexico. The average size of the region was more than 5,684 square miles from 2007 to 2012." [259]

[d] "Many parts of the U.S., including agricultural regions, are experiencing *groundwater depletion* (withdrawal exceeds recharge rate) at increasing rates." [260]

[e] "In 2011, farmers were reliant on income sources outside the farm to make up 83% of their household income, on average." [261]

[f] "Just 16¢ of every dollar *spent on food* in 2011 went back to the farm; in 1975, it was 40¢."[262]

In order to come up with a solution and to comprehend the benefits of this proposal macro-economic forecasting software could be used to simulate different ways of valuing the knowledge and features required to be a good farmer, the working conditions of the farmer, the best possible approach to addressing the needs of credit, supplies and knowledge for the rural population and the above-mentioned issues.

---

[258] ibid. Emphasis added.

[259] ibid. Emphasis added.

[260] ibid. Emphasis added.

[261] ibid. Emphasis added.

[262] ibid. Emphasis added.

THE FARMERS' SITUATION IN THE CONTEXT OF THE DESCRIPTIVE AND EXPERIMENTAL PHASE OF THE CYCLE OF SCIENTIFIC RESEARCH

Let us then continue with our example of the farmers using the using the Descriptive and Experimental Methods.

SUMMARY OF THE DESCRIPTIVE AND EXPERIMENTAL METHODS OF SCIENCE
Questions: With what being? With whom? With what type?
Essence: The Word of God. **To be:** animated and unanimated beings. **Forces:** power of mind, energy, capacities, potential, talents, vocations, arguments, concepts. **Categories:** genres, clusters, groups. Interacting entities within a system or a subsystem.
Laws of thermodynamics, labor **law** and tools **regulations**. Grammar **rules**. The Word of God is **Law** **Fruits of labor**: results, harvest, services, products and leftover materials or efforts
Powers: thought, reason and expression **To do:** movements: methods, processes, activities, abilities, skills, arts, technologies, mechanisms.
Relations of: Cause and effect, logic and reason.
Parameters and indicators of efficiency and productivity in achieving the results

Continuing with our example—and without intending that this be an exhaustive analysis—it is necessary to examine:

[1] The eating customs of humans and the resources used to feed livestock.

[2] The caloric, nutritional, industrial and health potential of the species being planted and of those that exist in the region that are not being planted

[3] The potentialities of the soil and the soil management techniques.

[4] The techniques employed in planting, cultivation, harvesting, packing and transporting, and the results being reaped from these.

Let us reflect on the sustainability of agribusiness:

After cars, the *food system uses more fossil fuel* than any other sector of the economy — 19 percent. And while the experts disagree about the exact amount, the way we feed ourselves *contributes more greenhouse gases to the atmosphere than anything else we do* — as much as 37 percent, according to one study. Whenever farmers clear land for crops and till the soil, large quantities of *carbon are released into the air.* But the 20th-century industrialization of agriculture has increased the amount of greenhouse gases *emitted by the food system* by an order of magnitude; chemical fertilizers (made from natural gas), pesticides (made from petroleum), farm machinery, modern food processing and packaging and transportation have together transformed a system that in 1940 *produced 2.3 calories of food energy for every calorie of fossil-fuel energy it used* into one that now *takes 10 calories of fossil-fuel energy to produce a single calorie of modern supermarket food.* Put another way, when we *eat* from the *industrial-food system*, we *are eating oil and spewing greenhouse gases*. This state of affairs appears

all the more absurd when you recall that every *calorie we eat is ultimately the product of photosynthesis — a process* based on *making food energy from sunshine.*[263]

Another source says: "Modern agriculture and the food system as a whole have developed a strong dependence on fossil energy; *7.3 units of (primarily) fossil energy are consumed for every unit of food energy produced.*"[264]

We must find a *system* where human labor is more intensive and more profitable for farmers but also a system where the symbiotic potential of plants of different *taxonomical families* becomes much more *efficient*, such as the three sisters' polyculture of maize, squash and beans. [265] [266]

[5] The potential damage that can be caused by plagues and the techniques farmers use to control plagues and diseases of plants and animals.

---

[263] Pollan, The Food Issue, Farmer in Chief. Emphasis added.

[264] Michigan University, Center for Sustainable Systems. Emphasis added.

[265] Postma, Complementarity in root architecture for nutrient uptake in ancient maize/bean and maize/bean/squash polycultures.

[266] The maize stalk serves as support for the climbing bean plant. The bean plant fixes nitrogen in the soil that benefits itself and the maize and the squash. The shade provided by the big leafs of the squash slow the evaporation process, which benefits the three sisters.

[6] The tools and equipment being employed and the ideal procedures that could be used in order to do it more efficiently.

With the same forecasting macro-economic software, examine the consequences of reversing the intentional displacement of rural populations, as captured in the following quote:

> For economic development to succeed in Africa in the next 50 years, African agriculture will have to change beyond recognition. *Production* will have to have increased massively, but also *labor productivity*, requiring a vast reduction in the proportion of the population engaged in agriculture and a large move out of rural areas. The paper questions how this can be squared with a continuing commitment to smallholder agriculture as the main route for growth in African agriculture and for poverty reduction. We question the evidence base for an exclusive focus on smallholders, and argue for a much more open-minded approach to different modes of *production*. To allow alternative modes and scale of *production* to emerge, new institutional and policy frameworks are required. A rush to establish 'mega-farms' with government discretionary allocation of vast tracts of land is unlikely to be the answer. Allowing a more dynamic agriculture to develop will require clear institutional frameworks, and not just a narrow focus on smallholders.[267]

[7] The techniques and methodologies employed by commercial establishments, private or governmental development agencies and those promoted by the educational system.

---

[267] Collier and Dercon, African Agriculture in 50 Years: Smallholders in a Rapidly Changing World? Emphasis added.

[8] Rain and residual water management, including disposal of organic matter.

These factors will help the population to define who they should consult in order to reach the change in their conception of the physical world and the potential of other beings (animate and inanimate) that they would like to utilize to reach their objectives. The population also has to make an effort to determine the parameters and indicators that may help them evaluate the efficiency of the community's products and services.

In order to think about agriculture as a way to "lay a foundation for *system*"[268], one has to examine the *systems of education and labor*. An *educational system* which revolves around agriculture should focus on monitoring and evaluating the *teaching of the methods of sciences* to the *youth* of the world, especially the *daughters and sons of the farmers*, so they can discover the *potential of biodiversity and its results in terms of generating employment*. For example: there are close to 300,000 *species of plants* and there are scores of different chemical compounds in a leaf of a plant, each one of them with different *potentialities*. The same thing applies to the possible combinations of those scores of different chemical compounds; each has its *potential for agriculture, nutrition, health or industrial uses*. "It has been estimated that well over 300,000 secondary metabolites exist, and it is thought that their primary function is to increase the likelihood of an organism's survival by repelling or attracting other organisms."[269] To understand the *potential of combinations* of 300,000 secondary metabolites, it is necessary to use a calculator such as http://stattrek.com/online-calculator/combinations-permutations.aspx. The *results for a set of 2 compounds* is 44,999,850,000 possible combinations;

---

[268] 'Abdu'l-Bahá, Foundations of World Unity, p. 39. Emphasis added.

[269] McMurry, John et al. Emphasis added.

for *a set of 4 compounds* is 3.3749325004125E+20; for *a set of 6 compounds* is 1.01244937595624E+30; and for *a set of 8 compounds* is 1.6270802736789E+39. How many different *uses* we can derive from the whole plant and its parts? How many beneficial *species* of insects, bacteria, viruses and fungus do we know? Because my work has been dedicated to the relentlessly persecuted Bahá'ís of Iran, I suggest this endeavor to be the continuation of the "Education is not a crime" initiative, and I would certainly love to be part of it.

In order to come up with a proposal, having *agriculture as a foundation for system,* use a macro-economy forecasting software to evaluate the *productivity* of the investment required in the *educational system*, to teach the *methods of science to the youth of the world to discover the potential of nature's biodiversity* and be at the forefront in the *generation of employment* when compared to the construction sector which today is considered the *motor* of the economy.

We should also monitor *employment* conditions when the principles of physics (*energy consumption per calorie produced*) help to correct the viability of *the agricultural system*.

The marginal *profitability for the farmers* and *salaries for those who work at the farm* should take into account the amount of *labor* required in comparison to those *who pack and/or transform the produce* and that of the *merchant*. Finally, but not less important, is to find out if those who are *self-employed* are being taken into account in the periodical government *report* about *unemployment* in the rural areas.

THE FARMERS' SITUATION IN THE CONTEXT OF THE PROPOSITIONAL PHASE OF THE CYCLE OF SCIENTIFIC RESEARCH

Let us then finish with our example of the farmers with the propositional method.

SUMMARY OF THE EXPLORATORY & PROPOSITIONAL METHODS OF SCIENCE
Question: With which option? when addressing problems, difficulties, issues and needs
Essence: Mercy. **Sources of Wisdom**: What history or experience, the diverse disciplines of science and the world religions say about the issue to be addressed **Kingdoms, Cycles and phases**
Laws: related to the inherent responsibilities of the assumed decision taking into account the effects in the other Kingdoms and in the Human Kingdom. **Goal**: Stewardship in the management of the trust
Powers: Intelligence, patience, self-control, discernment, compassion, empathy, volition **Decisions being made at the individual and collective level:** Governmental Institutions, Administration of private enterprises, NGOs, families or individuals. The human decisions in relation to the stages of a plan or phases of a cycle **The spontaneous differential response** to the stimuli received by the different kingdoms **The effects of the decisions made by humans in all kingdoms** **The developmental stage** of the individual, plant or animal **Monitoring and evaluating** the strategy and its stages and the cycle and its phases

> **Situations** in which the human kingdom is being challenged to choose or, in the case of the mineral, plant and animal kingdoms, the spontaneous differential response to the stimuli received.
>
> **Relationships:** are those that interconnect the sources of wisdom

> **Parameters and indicators of progress**: Such as level of respect for the individual to exercise his (her) free will, level of commitment, number of pledges, and number of recommendations or referrals

Finishing with our example of the farmers:

[1] From the statistics provided by the International Labor Organization in *Global Farm Worker Issues*, mentioned above, one can easily calculate that the number of self-employed farmers and unpaid family workers is 750 million. Therefore, 25 percent of the world's work force has been ignored in the calculations of *GDP*[270] and *unemployment rates, which are key parameters for governments to make their decisions*. It is not fair to make decisions only taking into account the 450 million who are hired agricultural workers. The contribution of these self-employed farmers and unpaid family workers should be taken into account in *GDP and the unemployment rates* for every region.

They are just part of the informal economy:

> Under this new definition, the informal economy is comprised of all forms of 'informal employment'— that is, employment without labour or social protection—both inside and outside informal enterprises, including both self-employment in small unregistered enterprises and wage employment in unprotected jobs. ...

---

[270] Gross Domestic Product

As part of economic restructuring and liberalization, there has been a fair amount of *deregulation*, particularly of financial and labour markets. *Deregulation* of labour markets is associated with the rise of informalization or 'flexible' labour markets. ...

Some countries include informal employment in agriculture in their estimates. This significantly increases the proportion of informal employment: from 83 per cent of non-agricultural employment to 93 percent of total employment in India; from 55 to 62 per cent in Mexico; and from 28 to 34 per cent in South Africa.[271]

Should we ask the Ministry of Labor or its equivalent in every country if he (she) includes farming work done by men and women in their own farms in the Gross Domestic Product (GDP)?

[2] We must decide which parameters and indicators will be used to control and evaluate the progress of the farmers' situation and to monitor and evaluate the protection of the ecosystem and our health. Ultimately the farmers should make more money than the agro-industry and the merchants. But, any agreement reached between these actors should take into account the principles of moderation and gradation. The change cannot be abrupt.

[3] The history of ancient civilizations such as Mesopotamia, Egypt, India and China provides a good idea of what farmers meant to the ruling classes as a background to understand what they mean to the current world leaders.

---

[271] Chen, Rethinking the Informal Economy: Linkages with the Formal Economy and the Formal Regulatory Environment. Emphasis added.

I want to encourage those interested in addressing this issue to study what happened when agriculture was discovered, how much farmers were taxed, who provided the army for the rulers to conquer new territories and who bore the burden of the construction of infrastructure. What caused those civilizations to collapse? (Toynbee, Arnold).

> Ian Johnson in an article entitled, Leaving the Land: China's Great Uprooting: Moving 250 Million Into Cities, says: "*China is pushing ahead with a sweeping plan to move 250 million rural residents* into newly constructed towns and cities over the next dozen years — a transformative event that could set off a new wave of growth or saddle the country with *problems for generations to come.*"

[4] We must evaluate the outcome for farmers and the community at large with a macro-economic forecasting software, in order to decide which administrative order and which authorities are strategic to be contacted for the success in the implementation of this proposal.

[5] We must monitor on a regular basis the level of hope of the farmers and their families about the viability of their farming businesses and the reasons why not to migrate towards the misery belts around big cities.

> Every man of *discernment*, while walking upon the earth, feeleth indeed abashed, inasmuch as he is fully aware that the thing which is the source of his prosperity, his wealth, his might, his exaltation, his advancement and power is, as ordained by God, the very earth which is trodden beneath the feet of all men. There can be no doubt that whoever is cognizant of this truth, is cleansed and sanctified from all pride, arrogance, and vainglory. Whatever hath been said hath come

from God. Unto this, He, verily, hath borne, and beareth now, witness, and He, in truth, is the All-Knowing, the All-Informed.[272]

******

It can be understood now that the scientific cycle of research consists of the exploratory phase at the beginning; followed by the explicative, the qualitative, the formative and the experimental phases in any sequence; and finally end with the propositional phase.

---

[272] Bahá'u'lláh, Epistle to the Son of the Wolf, p. 43. Emphasis added.

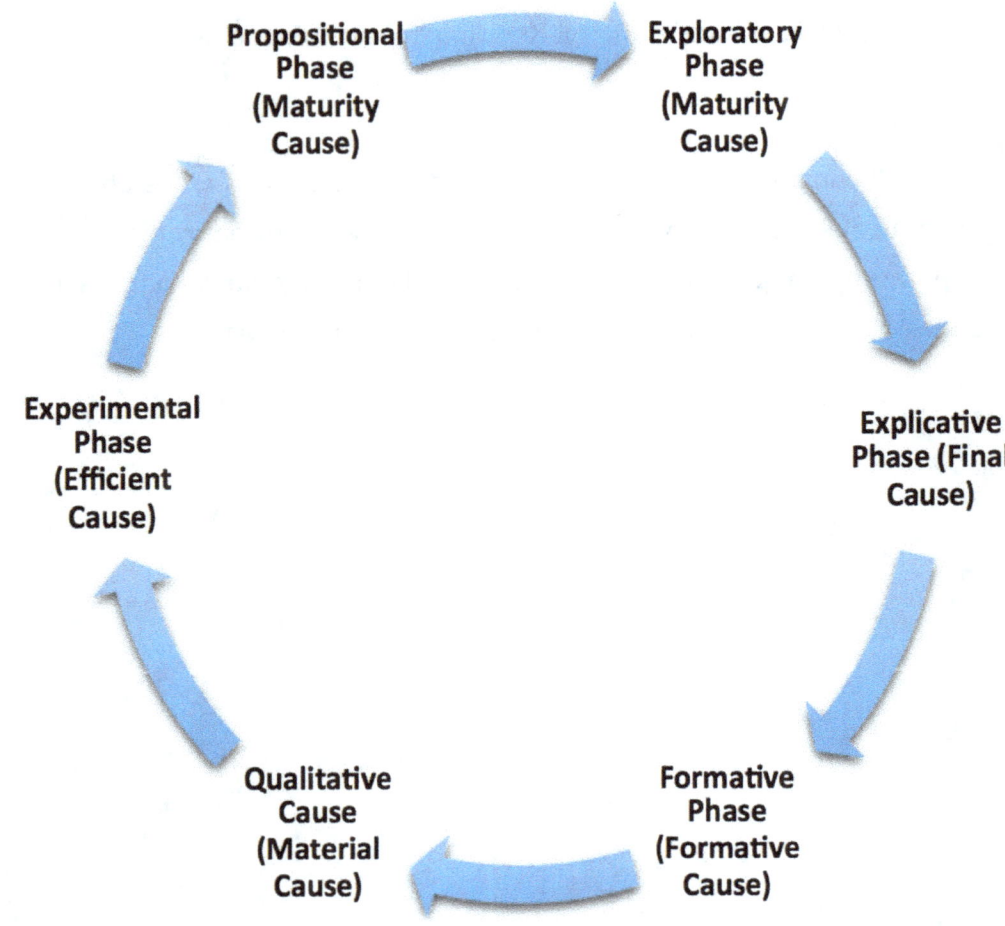

In conclusion, each method has its set of human faculties, an essence (also called a cause or origin) that is inherent to creation, associated laws, some moral values, a structure of relationships, a method, and parameters and indicators and is linked to at least one phase of the cycle of scientific research. The five causes are complementary and inherent to the essence of all things. Therefore, to reach a holistic description, none of the causes can be excluded.

Thomas Samuel Kuhn made several notable claims concerning the progress of scientific knowledge: that scientific fields undergo periodic 'paradigm shifts' rather than solely progressing in a linear and continuous way, and that these paradigm shifts open up new approaches to understanding what scientists would never have considered valid before; and that the notion of scientific truth, at any given moment, cannot be established solely by objective criteria but is defined by a consensus of a scientific community.[273]

It is not possible, any more, to perceive science and religion as non-overlapping magisteria, as conceived by Stephen Jay Gould. That is, science and religion each have "a legitimate magisterium, or domain of teaching authority,"[274] and these two domains do not overlap. With this conceptual framework, scholars and lay people can perceive that both Books, the Book of Revelation and the Book of Creation are wholly fused, interwoven in an inseparable holistic perspective.

Abdu'l-Bahá said:

> Religion and Science are inter-twined with each other and cannot be separated. These are the two wings with which humanity must fly. One wing is not enough. Every religion which does not concern itself with Science is mere tradition, and that is not the essential. Therefore science, education and civilization are most important necessities for the full religious life."[275]

---

[273] <https://en.wikipedia.org/wiki/Thomas_Kuhn>.

[274] Stephen Jay Gould. Nonoverlapping Magisteria.

[275] Abdu'l-Bahá, Abdu'l-Bahá in London, p. 28.

What would be the synergic effect when "science and religion, the two most potent forces in human life"[276] become reconciled?

---

[276] Shoghi Effendi in the Introduction of a book by Bahá'u'lláh, The Proclamation of Bahá'u'lláh, p. xi.

XI WORKS CITED

'Abdu'l-Bahá. *Abdu'l-Bahá in London*. Web. October 2016: Ocean Research Library.

_____ . *Divine Philosophy.* Web. June 2012
<http://bahairesearch.com/english/Baha%27i/Authoritative_Baha%27i/Abdu%27l-Baha/Divine_Philosophy.aspx>.

_____ . *Foundations of World Unity.* Web. October 2015
<http://reference.bahai.org/en/t/ab/>.

_____ . *Gems of Divine Mysteries.* Australia: Griffin Press, 2002. Print.

_____ . *Paris Talks: Addresses Given by 'Abdu'l-Bahá in Paris in 1911 in Writings and Utterances of 'Abdu'l-Bahá.* New Delhi, India: Bahá'í Publishing Trust, 2001. 695-795. Print.

_____ . *Selections from the Writings of 'Abdu'l-Bahá.* Great Britain: W & J Mackay Limited, Chatham, 1978. Print. Translated by a Committee at the Bahá'í World Centre and by Marzieh Gail. Haifa: Bahá'í World Centre, 1978.

_____ . *Some Answered Questions*. Web. October 2015
<http://reference.bahai.org/en/t/ab/>.

_____ . Tablet to Dr. Forel. Web. October 2015.
<http://www.bahai.org/library/authoritative-texts/downloads>.

_____ . *Tablets of 'Abdu'l-Bahá.* Web. June 2012
<http://reference.bahai.org/en/t/ab/TAB/tab-64.html>.

_____. *The Promulgation of Universal Peace* in Writings and Utterances of 'Abdu'l-Bahá. New Delhi, India: Bahá'í Publishing Trust, 2001. 797-1214. Print.

_____. *The Secret of Divine Civilization.* Wilmette, Ill.: Bahá'í Publishing Trust, 1990. Print.

_____. A *Traveller's Narrative.* Web. March 2014: Ocean Research Library.

Agronomists of the Potash and Phosphate Institute. *Phosphorus Uptake and Mycorrhizae - Partners in Plant Nutrition.* Agri-Briefs Agronomic News Items. Winter 1.999 No 2. Web. March 2014. <http://www.ipni.net/ppiweb/agbrief.nsf/5a4b8be72a35cd46852568d9001a18da/6daa3e4b2e5a3a4f852569040069b3bd!OpenDocument >.

Aristotle. *A Treatise on Government.* Trans. by William Ellis, A.M. The Project Gutenberg EBook of Politics, by Aristotle, 2009.Web. June 2012 <http://www.gutenberg.org/files/6762/6762-h/6762-h.htm#2HCH0084>.

_____. *The Metaphysics.* Web. Nov 2015. <http://classics.mit.edu/Aristotle/metaphysics.mb.txt>.

Bahá'u'lláh. *Epistle to the Son of the Wolf.* Trans. Shoghi Effendi. Wilmette, Ill.: Bahá'í Publishing Trust, 1962. Print.

_____. *Gleanings from the Writings of Bahá'u'lláh.* Trans. Shoghi Effendi. Wilmette, Ill.: Bahá'í Publishing Trust, 1976. Print.

_____. *Prayers and Meditations by Bahá'u'lláh.* Trans. Shoghi Effendi. Wilmette, Ill.: Bahá'í Publishing Trust, 1974. Print.

_____ . *Tablets of Bahá'u'lláh*. Great Britain: W & J Mackay Limited, Chatham, 1978. Print.

_____ . *Tablets of Bahá'u'lláh Revealed After the Kitáb-i-Aqdas*. Web. October 2015 < http://reference.bahai.org>.

_____ . *The Hidden Words of Bahá'u'lláh*. Trans. Shoghi Effendi. Wilmette, Ill.: Bahá'í Publishing Trust, 1980. Print.

_____ . *The Kitáb-i-Aqdas*: The Most Holy Book. Ann Arbor, Michigan: Edward Brothers, 1992. Print.

_____ . *The Proclamation of Bahá'u'lláh*. Web. October 2016: Ocean Research Library.

_____ . *The Seven Valleys and the Four Valleys*. Trans. Marzieh Gail. Wilmette, Ill.: Bahá'í Publishing Trust, 1991. Print.

_____ . *The Tabernacle of Unity*. Web. March 2017: Ocean Research Library.

Bahá'u'lláh, Abdu'l-Bahá, Shoghi Effendi and Universal House of Justice. *Conservation of the Earth's Resources* in *Compilation of Compilations*. Volume 1, 1991. Web. June 2012 <http://bahai-library.com/compilation_conservation_resources>.

Barrett, Christopher B., Bachke, Maren E., Bellemare, Marc F., Michelson, Hope C., Narayanan, Sudha and Walker, Thomas F. *Smallholder Participation in Agricultural Value Chains: Comparative Evidence from Three Continents*. December 2010 revision. Web. January 2018 < barrett.dyson.cornell.edu/.../SmallholderMarketParticipationDec2010Submission.pdf>.

Barry, Ellen. *After Farmers Commit Suicide, Debts Fall on Families in India* in New York Times 22 February 2014. Web. March 2014 <http://www.nytimes.com/2014/02/23/world/asia/after-farmers-commit-suicide-debts-fall-on-families-in-india.html>.

Biology Online, Web. July 2017 <http://www.biology-online.org/>.

Born, Max, Nobel Prize-winning physicist, qtd. in Gerald Holton's *Thematic Origins of Scientific Thought*. Web. October 2015 < https://books.google.com/books?id=vAv5YmGWosoC&pg=PA7&dq=Born,+Max,+Thematic+Origins+of+Scientific+Thought+scouts+erect&hl=en&sa=X&ved=0ahUKEwjS3oihwqLRAhXGZCYKHcFdBcAQ6AEIHDAA#v=onepage&q=Born%2C%20Max%2C%20Thematic%20Origins%20of%20Scientific%20Thought%20scouts%20erect&f=false>.

Brockett, Charles D., *Land Power and Poverty. Agrarian Transformation and Political Conflict in Central America*. Westview Press, 1998, Print.

Business Dictionary. Web. July 2017 <http://www.businessdictionary.com/>.

Center for Sustainable Systems, Michigan University. *U.S. Food System*. Web. March 2014 <http://css.snre.umich.edu/css_doc/CSS01-06.pdf>.

Chen, Martha Alter. *Rethinking the Informal Economy: Linkages with the Formal Economy and the Formal Regulatory Environment*. DESA Working Paper No. 46 July 2007

Collier, Paul, and Stefan Dercon. *African Agriculture in 50 Years: Smallholders in a Rapidly Changing World?*, Elsevier . Web. December 2013 <http://www.sciencedirect.com/science/article/pii/S0305750X13002131>.

Conant, James Bryant. qtd. in *Science and Common Sense.* Web. October 2015 < http://www.gly.uga.edu/railsback/1122sciencedefns.html >.

Davidson, Thomas. *Aristotle and Ancient Educational Ideals.* Web, June 2017 <http://www.gutenberg.org/files/40552/40552-h/40552-h.htm>.

Dixon, John, Gulliver, Aidan and David Gibbon. *Farming Systems and Poverty: improving farmers' livelihoods in a changing world.* Web. March 2014 <http://www.fao.org/farmingsystems/>.

Duque, Leonardo, *En la Encrucijada: Una Perspectiva Nueva: En Honor a las Mujeres del Mundo.* Cali: Editorial Cargraphics, 1999. Print.

Esslemont J. E., Bahá'u'lláh and the New Era. Wilmette, Ill.: Bahá'í Publishing Trust, 1978. Print.

Elitzak, H. *Food Cost Review, 1950-97.* USDA, Agricultural Economic Report 780. 1999. qtd. in Center for Sustainable Systems. Web. March 2014 <http://css.snre.umich.edu/css_doc/CSS01-06.pdf>.

Encyclopedia Britannica. Web July 2017 <https://www.britannica.com/topic/European-Community-European-economic-association>.

Feynman, Richard P., Nobel prize winning physicist. *Religion is a culture of faith; science is a culture of doubt.* Web. October 2015 <http://www.gly.uga.edu/railsback/1122sciencedefns.html>.

_____. *The Pleasure of Finding Things Out* 1999. Web. October 2015 <http://www.gly.uga.edu/railsback/1122sciencedefns.html >.

Gould, Stephen J., *Nonoverlapping Magisteria.* 1997. Web. December 2016 <http://www.stephenjaygould.org/library/gould_noma.html>.

Hatcher, William S. *Logic and Logos.* Oxford: George Ronald, 1990. Print.

Hauser, Philip M. (1909-1994), Demographer and Census Expert, qtd. in Theodore Berland's *The Scientific Life.* Web. June 2012 <http://www.gly.uga.edu/railsback/1122sciencedefns.html>.

Heller, Martin C. and Keoleian Gregory A. *Life Cycle-Based Sustainability Indicators for Assessment of the U.S. Food System*, The University of Michigan Center for Sustainable Systems, 2000. qtd. in Center for Sustainable Systems. Web. March 2014 <http://css.snre.umich.edu/css_doc/CSS01-06.pdf>.

Hershberg, James G. *Harvard to Hiroshima and the Making of the Nuclear Age.* 1.993. Web. January 2017 <https://books.google.com/books?id=fqhzrXn1RE0C&pg=PA410&dq=%22in+his+controversial+The+Structure+of+Scientific+Revolutions+(1962)%22&hl=en&sa=X&ved=0ahUKEwjQ94D19KHRAhUJOyYKHc6qBoMQ6AEIHDAA#v=onepage&q=%22in%20his%20controversial%20The%20Structure%20of%20Scientific%20Revolutions%20(1962)%22&f=false>.

Heussner, Ki Mae. ABC News. Web. April 2015
<http://abcnews.go.com/WN/Technology/stephen-hawking-religion-science-win/story?id=10830164http://abcnews.go.com/WN/Technology/stephen-hawking-religion-science-win/story?id=10830164>.

Hofstede, Geert. *Culture's Consequences: International Differences in Work Related Values.* Beverly Hills CA: Sage Publications, 1980.

Hornby, Helen. comp. *Lights of Guidance: A Bahá'í Reference File.* First online edition, 2006. Web. June 2012 <http://bahai-library.com/hornby_lights_guidance_2.html&chapter=2>.

_____. Compilations, *Lights of Guidance*. Web. March 2014: Ocean Research Library.

Huitt, William. *What Is a Human Being and Why Is Education Necessary.* May 2017. Web. August 2017 <http://www.edpsycinteractive.org/topics/intro/human.html>.

_____ *Citizenship. Cosmic-Citizenship.* 2015. Web. August 2017 <http://www.cosmic-citizenship.org/>.

International Labor Organization (ILO). *Agriculture; plantations; other rural sectors.* Web. January 2015 <http://ilo.org/global/industries-and-sectors/agriculture-plantations-other-rural-sectors/lang--en/index.htm>.

_____ ILO Statement to the 56th Commission on the Status of Women. *Adoption of international labour standards key to supporting rural women.* New York March 2012. Web. July 2017

<http://www.ilo.org/newyork/at-the-un/commission-on-the-status-of-women/WCMS_209379/lang--en/index.htm>.

_____ *Global Farm Worker Issues.* October 2003 Volume 9 Number 4.Web. February 2014 <http://migration.ucdavis.edu/rmn/more.php?id=785_0_5_0>.

Jammer, Max. *Einstein and Religion.* Princeton University Press, 1999. Web. December 2016 <file:///C:/Users/Leonardo%20Duque/Dropbox/Downloads/s6681(1).pdf>.

Johnson, Ian. *Leaving the Land: China's Great Uprooting: Moving 250 Million Into Cities.* New York Times June 2013. Web. March 2014 <http://www.nytimes.com/2013/06/16/world/asia/chinas-great-uprooting-moving-250-million-into-cities.html?pagewanted=all>.

Joseph, Enamuthu, *Global Nutrition and Development.* Center for Human Resources, State University of New York at Plattsburgh. 1996.

Kemerling, Garth. *A Dictionary of Philosophical Terms and Names.* Web. 18 June 2012 <http://www.philosophypages.com/dy/index.htm>.

Kessler, H. G. *The Diary of a Cosmopolitan*, (London: Weidenfeld and Nicolson, 1971), p.157; quoted in *Einstein and Religion* by Max Jammer (Princeton University Press, 1999) pp. 39-40.Web. April 2015 <http://einsteinandreligion.com/religioncomments.html>.

Kluge, Ian. *The Aristotelian Substratum of the Baha'i Writings.* Web. July 2017 < https://www.bahaiphilosophy.com/the-aristotelian-substratum-1-.html>.

Konikow, L. *Groundwater depletion in the United States (1900-2008).* U.S. Geological Survey (USGS) Scientific Investigations Report. 2013. qtd. in Center for Sustainable Systems. Web. March 2014 <http://css.snre.umich.edu/css_doc/CSS01-06.pdf>.

Kwelagobe, Mpule K. *Commentary - Investing in Rural Women: Closing the Gender Gap in African Agriculture.* Web. February 2014 <http://globalfoodforthought.typepad.com/global-food-for-thought/2013/10/commentary-investing-in-rural-women-closing-the-gender-gap-in-african-agriculture-1.html>.

McCarthy, Eugene M. *Online Biology Dictionary.* Web. July 2017 <http://www.macroevolution.net/biology-dictionary.html>.

McKeon, Richard. *The Basic Works of Aristotle.* New York: Random House, 1941. Print.

McMurry John. *Organic Chemistry with Biological Applications.* Cengage Learning, Apr 14, 2014 - Science - 672 pages. Web. https://books.google.com/books?id=KDIeCgAAQBAJ&pg=PP4&dq=Organic+Chemistry+with+Biological+Applications++John+McMurry&hl=en&sa=X&ved=0CCcQ6AEwAGoVChMIybrm2ubCxwIViXc-Ch2pBgqq#v=onepage&q=Organic%20Chemistry%20with%20Biological%20Applications%20%20John%20McMurry&f=false

Martorell, Reynaldo. *Undernutrition During Pregnancy and Early Childhood: Consequences for Cognitive and Behavioral Development.* Elsevier Science B.V. 1.997.

Merriam-Webster Dictionary. Web. October 2012. <www.Merriam-Webster.com>.

Ministry of Foreign Affairs of Denmark. August 2010. *Gender and Value Chain Development.* January 2018 < http://www.netpublikationer.dk/um/10511/html/entire_publication.htm#Section4.1 >.

Mora, Camilo, Derek P. Tittensor, SinaAdl, Alastair G. B. Simpson and Boris Worm. *How Many Species on Earth and in the Ocean?* Web. April 2014 <http://www.plosbiology.org/article/info%3Adoi%2F10.1371%2Fjournal.pbio.1001127>.

Morgenstern, Julian. *A Jewish Interpretation of the Book of Genesis.* Web. October 2015 <https://books.google.com/books?id=SJ4sAAAAYAAJ&pg=PA74&lpg=PA74&dq=%22Eye+for+eye,+and+tooth+for+tooth%22+nomad&source=bl&ots=xjGsfD34-p&sig=eEnN9poHgfsqLiVq-cscorzoOQo&hl=en&sa=X&ved=0CB0Q6AEwAGoVChMIjuqzjqrryAIVRzkmCh14HAYL#v=onepage&q=%22Eye%20for%20eye%2C%20and%20tooth%20for%20tooth%22%20nomad&f=false>.

National Oceanic and Atmospheric Administration (NOAA). "NOAA Scientists: Midwest drought brings fourth smallest Gulf of Mexico 'Dead Zone' since 1985. 2008. qtd. in Center for Sustainable Systems. Web. March 2014 <http://css.snre.umich.edu/css_doc/CSS01-06.pdf>.

Plato. *Crito: or, the Duty of a Citizen.* Project Gutenberg's EBook of Apology, Crito, and Phaedo of Socrates, by Plato 2004.Web. June 2012<http://www.gutenberg.org/files/13726/13726-h/13726-h.htm - crito_or_the_duty_of_a_citizen>.

_____ . *Sophist.* Trans. Benjamin Jowett. The Project Gutenberg EBook of Sophist, by Plato, 2008.Web. June 2012 <http://www.gutenberg.org/files/1735/1735-h/1735-h.htm>.

_____ . *Symposium.* Trans. Benjamin Jowett. The Project Gutenberg EBook of Symposium, by Plato, 2008.Web. June 2012 <http://www.gutenberg.org/files/1600/1600-h/1600-h.htm>.

_____ . *Theaetetus.* Trans. Benjamin Jowett. The Project Gutenberg EBook of Theaetetus, by Plato 2008.Web. June 2012 <http://www.gutenberg.org/files/1726/1726-h/1726-h.htm>.

Plotinus. *The Six Enneads.* Trans. by Stephen Mackenna and B. S. Page.Web. June 2012 <http://classics.mit.edu/Plotinus/enneads.3.third.html>.

Popper, Karl. *Truth and the growth of knowledge.* 1962. Web. March 2015 <http://books.google.com.co/books?hl=en&lr=&id=YzvKJ-2nJn4C&oi=fnd&pg=PA285&dq=%22Popper,+Karl%22+1962+falsifiability&ots=UVEzPyo3Ao&sig=r2DUgYQAM7WlJ-36rgqi_9nsOMA#v=onepage&q=refutability&f=false

Ridley, Matt. *Genome: the autobiography of a species in 23 chapters*, p. 271.1999. Web. October 2015 <http://www.gly.uga.edu/railsback/1122sciencedefns.html>.

Robin, León. *El Pensamiento Griego y los Orígenes del Espíritu Científico. La Evolución de la Humanidad.* Trans. by the author. Enciclopedia Uteha, Vol. 14. Unión Tipográfica Editorial Hispano Americana, México, 1.962. Print.

Satz, Mario. *Ecología y Mitología.* Trans. by the author. Revista Nueva Conciencia. Barcelona, España. Integral Ed., 1991. Print.

*Shayast-na-Shayast 13:29.*Web: January 2013 <http://www.thesynthesizer.org/golden.html>.

Schnitzer, Stefan A., Klironomos JN, Hillerislambers J, Kinkel LL, Reich PB, Xiao K, Rillig MC, Sikes BA, Callaway RM, Mangan SA, van Nes EH, Scheffer M. *Soil microbes drive the classic plant diversity-productivity pattern.* Ecology. 2011 Feb;92(2):296-303. Web. October 2015 <http://www.ncbi.nlm.nih.gov/pubmed/21618909>

Shoghi Effendi. *Bahá'í Administration.* Wilmette, Illinois: Bahá'í Publishing Trust, 1960. Print.

_____. *Directives from the Guardian.* Web. March 2014: Ocean Research Library.

_____. Summary Statement - 1947, Special UN Committee on Palestine. Web. January 2015: Ocean Research Library.

_____ . *The Advent of Divine Justice.* Web. October 2015 <http://reference.bahai.org/en/t/se/ADJ/adj- 2.html>.

_____. *The World Order of Bahá'u'lláh - Selected Letters*" in Compilations, The Compilation of Compilations vol. I.

Star of the West. Web. July 2017: Ocean Research Library.

Stenmark, Mikael. *Rationality and Different Conceptions of Science* quoted in Wentzel van Huyssteen on Rationality in Science and Theology. A Discussion Note on F. LeRon Shults (ed.). The Evolution of Rationality. Interdisciplinary Essays in Honor of J. Wentzel van

Huyssteen. Grand Rapids, MI. /Cambridge, U.K.: Eerdmans,2006.ArsDisputandiVolume 9 (2009). Web. December 2016 <https://books.google.com/books?id=r3y4FdVCQ1sC&pg=PA51&lpg=PA51&dq=%E2%80%9CScience+should+be+nonresponsible+in+the+sense%22&source=bl&ots=AXA0f3sijJ&sig=sEwiIiwS0tLikMPUvae2Zig_1AY&hl=en&sa=X&ved=0ahUKEwj217Pp1fbQAhXmg1QKHT68AiQQ6AEIHDAA#v=onepage&q=%E2%80%9CScience%20should%20be%20non-responsible%20in%20the%20sense%22&f=false>.

Streiff, Jefrey. *The Golden Rule*. Web. December 2016
<http://www.goldenruleart.com/>.

Stice, Kyle and Sale, Andrew. *Aruligo Pineapple Value Chain – Mapping Report.* November 2008. Web. January 2018 < www.pacificfarmers.com/wp.../06/Solomon-Islands-Pineapple-Mapping-report.pdf>.

Sun, Joseph C., Lopez-Verges, Sandra, Kim, Charles C., DeRisi, Joseph L. and Lewis L. Lanier. *NK Cells and Immune "Memory".* Web. May 2014 <http://www.jimmunol.org/content/186/4/1891.short>.

Taherzadeh, Adib. *The Revelation of Bahá'u'lláh* v 4. Oxford: George Ronald, Publisher, 1988. Print.

The Bible, Luke 6:30-36. Web. January 2013
<http://www.goldenruleart.com/>.

The Mahabharata. Web. January 2013 <http://www.goldenruleart.com/>.

*The Talmud, Shabbat 31a.*Web. January 2013
<http://www.thesynthesizer.org/golden.html>.

The Universal House of Justice. *The Promise of World Peace.* New Delhi, India: Bahá'í Publishing Trust, 1992. Print.

_____ . *Consultation: A Compilation.* Wilmette, Illinois: Bahá'í Publishing Trust, 1980.Print.

_____ . *A Compilation on Women.* Compiled by the Research Department of the Universal House of Justice, Bahá'í World Centre, January 1986. Web. October 2015 <http://reference.bahai.org/en/t/c/CW/>.

Traoré, Doussou. *Cocoa and Coffee Value Chains in West and Central Africa: Constraints and Options for Revenue-Raising Diversification.* Food and Agriculture Organization

of the United Nations. February 2009. Web. January 2018 < www.fao.org/fileadmin/templates/est/.../FAO_AAACP_Paper_Series_No_3_1_.pdf >.

Toynbee, Arnold. *Estudio de la Historia.* Compendio. Madrid: Alianza Editorial. 1981. Print.

U.S. Environmental Protection Agency (EPA). *Pesticide Industry Sales and Usage: 2006 and 2007 Market Estimates.* 2011. qtd. in Center for Sustainable Systems. Web. March 2014 <http://css.snre.umich.edu/css_doc/CSS01-06.pdf>.

USDA, Economic Research Service. *Characteristics of principal farm operator households, by gross farm sales, 2011.* 2012. qtd. in

Center for Sustainable Systems. Web. March 2014
<http://css.snre.umich.edu/css_doc/CSS01-06.pdf>.

_____. *Food Dollar Series*. 2013. Web. March 2014
<http://css.snre.umich.edu/css_doc/CSS01-06.pdf>.

USDA, National Resources Conservation Service. *2007 National Resources Inventory*. 2009. qtd. in Center for Sustainable Systems. Web. March 2014
<http://css.snre.umich.edu/css_doc/CSS01-06.pdf>.

Walsh, Roger. *El Compromiso con el Planeta.* Revista Nueva Conciencia. p. 82. Trans. by the author

_____. *Staying Alive: the psychology of human survival.* Boulder, Colorado: Shambhala Publications Inc., 1984. Print.

Wikipedia. *Anthropocentrism*. Web. December 2016
<http://en.wikipedia.org/wiki/Anthropocentrism>.

_____. *Korsakoff Syndrome*. Web. March 2015
<http://en.wikipedia.org/wiki/Korsakoff%27s_syndrome>.

ISBN 978-1-7320081-2-0

www.ingramcontent.com/pod-product-compliance
Lightning Source LLC
Chambersburg PA
CBHW051403070526
44584CB00023B/3266